Lecture Notes in Computer Science 16450

The series Lecture Notes in Computer Science (LNCS), including its subseries Lecture Notes in Artificial Intelligence (LNAI) and Lecture Notes in Bioinformatics (LNBI), has established itself as a medium for the publication of new developments in computer science and information technology research, teaching, and education.

LNCS enjoys close cooperation with the computer science R & D community, the series counts many renowned academics among its volume editors and paper authors, and collaborates with prestigious societies. Its mission is to serve this international community by providing an invaluable service, mainly focused on the publication of conference and workshop proceedings and postproceedings. LNCS commenced publication in 1973.

Irina Mihaela Ciortan · Daniele Ferdani ·
Jon Yngve Hardeberg · Sophia Sotiropoulou ·
Manuele Veggi
Editors

Chromatic Visions

Exploring Colour in Art, Archaeology and Digital Realities, Part II

Springer

Editors
Irina Mihaela Ciortan
Norwegian University of Science and Technology
Gjøvik, Norway

Jon Yngve Hardeberg
Norwegian University of Science and Technology
Gjøvik, Norway

Manuele Veggi
University of Rome "La Sapienza"/National
Research Council – Institute of Heritage Science
Sesto Fiorentino, Italy

Daniele Ferdani
National Research Council – Institute of Heritage
Science
Montelibretti, Italy

Sophia Sotiropoulou
Hellenic Open University/Foundation for Research
and Technology Hellas
Patra, Greece

ISSN 0302-9743 ISSN 1611-3349 (electronic)
Lecture Notes in Computer Science
ISBN 978-3-032-17699-8 ISBN 978-3-032-17700-1 (eBook)
https://doi.org/10.1007/978-3-032-17700-1

This Springer imprint is published by the registered company Springer Nature Switzerland AG
The registered company address is: Gewerbestrasse 11, 6330 Cham, Switzerland

Preface

The PERCEIVE project (Perceptive Enhanced Realities of Coloured Collections through AI and Virtual Experiences), funded by the European Commission under Horizon Europe (Grant Agreement No. 101061157) [2], explores how digital technologies can advance the study, reconstruction, and communication of colour in cultural heritage. By combining heritage science, artificial intelligence, and interactive design, PERCEIVE seeks to create new ways to perceive, preserve, and share coloured collections across both physical and virtual environments.[1]

The project's interdisciplinary framework is articulated through five core scenarios, each representing a specific typology of coloured collection: (1) lost polychromy in ancient sculpture and architecture, with case studies taken from the National Archaeological Museum of Naples (MANN); (2) colour alteration in paintings and works on paper, focussing on Munch's *Scream* and *Old Man in Warnemünde* (Munch Museum, Oslo), on Cezanne's *Road in Provence* (AIC), and on the frescoes from the Temple of Isis in Pompeii (MANN); (3) fading of textiles and (4) degradation of colour in early photographs and film, with case studies from the collection of the Victoria & Albert Museum; and (5) authentic display of born-digital artworks, in collaboration with HSLU (Switzerland).

These scenarios guide scientific investigations into material change, as well as the design of digital tools, services, and experiences that connect researchers, curators, and citizens in renewed relationships of understanding and care for colour in cultural heritage [3].

The first volume of *Chromatic Visions* gathered the theoretical foundations developed during the initial phase of PERCEIVE [4]. It outlined the project's contributions on colour science and polychromy reconstruction, as well as its conceptual pillars (Authenticity, Sense of Care, and Participation) guiding interactive experience design. Drawing on philosophy, psychology, conservation science, and human–computer interaction, these chapters established a shared vocabulary and design rationale to interpret the fragility of colour and to envision new, meaningful ways to engage with heritage collections.

While the first volume established the conceptual and theoretical framework of the PERCEIVE project, this second book responds to a complementary need: the translation of those principles into practice. It addresses a growing demand in the field of cultural heritage for operational methods capable of linking scientific analysis, digital reconstruction, and audience engagement within a coherent and reproducible framework.

---

[1] The research consortium, coordinated by Italian National Research Council - Institute of Heritage Science (CNR - ISPC), is composed of seven participants and four partnersFoundation for Research and Technology - Hellas (FORTH, Greece); Anamnesia and IMKI (France); Norwegian University of Science and Technology (NTNU); Fraunhofer Institute for Computer Graphics Research (IGD, Germany); National Archaeological Museum of Naples (MANN, Italy); Munch Museum, Oslo (Norway); Art Institute of Chicago (AIC, USA); Overlay, Inc. (USA); Lucerne University of Applied Sciences and Arts (HSLU, Switzerland); Victoria & Albert Museum, London (V&A, UK).

The value of this volume lies in its integrative approach. While previous research has often explored colour reconstruction or digital museology as separate domains, *Chromatic Visions* II demonstrates how scientific models, interoperable infrastructures, and participatory design can coexist within a single methodological continuum. In this sense, it not only documents project outcomes, but also provides a reference model for how interdisciplinary research can effectively inform conservation strategies and cultural mediation practices.

This second volume presents the results and applications deriving from those theoretical contributions. It documents the suite of technological tools, design resources, and interactive prototypes co-developed by the PERCEIVE consortium to translate research on colour, perception, and care into practice. It is structured into three chapters, corresponding to complementary levels of this process—from scientific modelling to platform integration and experiential design. As for the first part, all contributions have been revised internally and externally by the project's external scientific board, anonymously.

Chapter 1, *Methods and Workflows for Colour Change Modelling and Reconstruction*, discusses the PERCEIVE research on coloured collections across the five scenarios, detailing how analytical data and mock-up studies informed the development of new models for simulating and predicting colour change. It describes the workflow through which heritage scientists use data-driven and AI-based methods to reconstruct lost or altered appearances and model the impact of degradation factors over time. These studies lay the scientific groundwork for subsequent tools and visualisation services.

Chapter 2, *Tools and Services for Colour Change Analysis and Sharing of Knowledge*, introduces the digital tools, services, and knowledge infrastructure created by the project, including the *PERCEIVE Platform* and *Colour Knowledge Repository*. These include the *MuLaX Web3D* viewer, which enables the multi-layer exploration of diagnostic data, *StyleShade3D*, which provides AI-assisted rendering of coloured surfaces, the *Light Damage Estimator*, which supports preventive conservation, and image-based tools such as *Deep Degreening* and *Text2Autochromes*, which facilitate the processing of historical photography. All of these resources are integrated through a FAIR-compliant repository, which ensures open access to, and the long-term re-use and interoperability of, colour-related data.

Chapter 3, *Design Resources to Enhance Access to Coloured Collections through Interactive Experiences*, explores how the theoretical principles of authenticity, care, and participation are translated into tangible visitor experiences. It introduces the *PERCEIVE Design Toolbox*, a suite including the *CoDesign Tool*, *Prototyper*, and *Experiment Planner*, which supports the ideation, prototyping, and evaluation of digital heritage applications. The chapter also presents the project's main demonstrators and prototypes, showing how scientific knowledge on colour can become accessible and emotionally engaging for diverse audiences. Among these are the exhibition *Perceive Isis' Colours*, featuring the authenticity prototype *The Gifts of Isis* and the caring prototype *Echoes of Loss*, as well as *Tiny Conservators*, a cooperative educational game, and the *Open Space Museum*, which extends these experiences into the urban environment.

This second volume completes the trajectory initiated in *Chromatic Visions* Volume I, taking us from conceptual exploration to concrete implementation. The outputs presented here demonstrate how digital research and design can bridge science, art, and public participation to re-establish the chromatic dimension of cultural heritage and foster a shared sense of responsibility and engagement.

Beyond the specific contributions of each chapter, this volume collectively represents a significant advancement in the fields of conservation and digital restoration. By framing colour restoration as a scientific field and a means of cultural communication, this volume seeks to expand the traditional notion of restoration, shifting the emphasis from physical intervention to a process involving reconstruction, interpretation, and knowledge sharing. These processes are underpinned by transparent data and reproducible workflows.

The tools and methodologies presented here contribute to an evolving culture of restoration that integrates scientific evidence, digital innovation, and collaborative practice, combining analytical precision with accessibility and long-term knowledge sustainability. They are not just technologically novel, but also interconnect in original ways: predictive models inform visual tools; repositories ensure data traceability and reuse, and immersive environments transform expert knowledge into shared understanding. In doing so, *Chromatic Visions* Vol. 2 provides a concrete reference for sustainable, collaborative, and ethically aware practices in heritage preservation.

Looking ahead, the methodologies, tools, and platforms described in this book open new directions for research and professional practice. Their modular and open design makes them adaptable and reusable beyond the *PERCEIVE* case studies, offering potential applications in areas such as preventive conservation, museum planning, digital archiving, education, and creative reuse. As these resources evolve, they encourage the heritage community to adopt more integrated, participatory, and evidence-based approaches to the study of colour and material change.

This process will continue within the Horizon Europe project *COLOURS* [1], Collaborative On-cloud Lab for the conservation and digital restoration of ColOUred heritage collectionS (Grant agreement ID: 101233413), which builds upon and extends PERCEIVE's outcomes to further advance the modelling, visualisation, and communication of colour in cultural heritage.

In this perspective, this volume is not a conclusion, but a step towards a broader ecosystem of shared infrastructures, where science, design, and the care of cultural heritage can develop together in transparent and sustainable ways.

Project External Scientific Board. Francesca Casadio, Hartmut Kutzke, Catlin Langford, Paolo Liverani, Rebecca Pedrazzi, Michael Weinmann, Giovanni Verri, Peter Nussbaum. The editors thank Hartmut Kutzke, Paolo Liverani, and Peter Nussbaum for their support as peer reviewers of this volume.

Acknowledgments. This study was funded by the European Union (grant number 101061157, Project PERCEIVE). The views and opinions expressed in this book are however those of the author(s) only and do not necessarily reflect those of the European Union or the European REA.

Disclosure of Interests. The authors have no competing interests to declare that are relevant to the content of this article.

Irina Mihaela Ciortan
Daniele Ferdani
Jon Yngve Hardeberg
Sophia Sotiropoulou
Manuele Veggi

References

1. COLOURS: Collaborative On-cloud Lab for the conservation and digital restoration of ColOUred heritage collectionS (2025). https://doi.org/10.3030/101233413, European Commission, CORDIS
2. PERCEIVEPerceptive Enhanced Realities of Colored collEctions through AI and Virtual Experiences (2023). https://doi.org/10.3030/101061157, European Commission, CORDIS
3. Pescarin, S., Aga, B., Arteaga Kiyomoto, I.Y., Barandoni, C., Bonifazi, F., Boracchi, B., Burgio, L., Cerato, I., Ciortan, I., Clay, A., Fanini, B., Fedon, G.C., Ferdani, D., Graf, H., Hardeberg, J.Y., Magrini, D., Massidda, M., Panteleris, P., Papadopoulos, G., Pitikakis, M., Sandu, I.C.A., Sinha, S.N., Siozos, P., Sotiropoulou, S., Spotti, S., Stavroulakis, P., Travaglini, L., Trumpy, G., Veggi, M.Perceptive Enhanced Realities of Coloured collEctions through AI and Virtual Experiences. In: Campana, S., Ferdani, D., Graf, H., Guidi, G., Hegarty, Z., Pescarin, S., Remondino, F. (eds.) Digital Heritage 2025. The Eurographics Association (2025). https://doi.org/10.2312/dh.20253350
4. Pescarin, S., Barandoni, C., Clay, A., Papadopoulos, G., Sandu, I.C.A. (eds.): Chromatic visions. Exploring Colour in Art, Archaeology and Digital Realities, Part I. Lecture Notes in Computer Science, vol. 15750, Springer Nature, Cham, Switzerland (Dec. 2025), 10.1007/978-3-032-07792-9

Organization

Scientific Committee

Francesca Casadio	Art Institute of Chicago, USA
Hartmut Kutzke	Museum of Cultural History, Norway
Catlin Langford	Centre for Contemporary Photography, Australia
Paolo Liverani	University of Florence, Italy
Roberta Pedrazzi	"NotiziArte", Italy
Michael Weinmann	Delft University of Technology, The Netherlands
Giovanni Verri	Art Institute of Chicago, USA
Peter Nussbaum	NTNU, Norway

PERCEIVE Coordinator

Sofia Pescarin	CNR ISPC, Italy

PERCEIVE Scientific Board

Cristiana Barandoni	CNR ISPC, Italy
Lucia Burgio	V&A, UK
Ivana Cerato	CNR ISPC, Italy
Arthur Clay	HSLU, Switzerland
Brenda Doherty	CNR SCITEC, Italy
Daniele Ferdani	CNR ISPC, Italy
Maria Lucia Giacco	MANN Museum, Italy
Holger Graf	NTNU, Norway
Catlin Langford	Centre for Contemporary Photography, Australia
Donata Magrini	CNR ISPC, Italy
Florent Michel	IMKI, France
Letizia Monico	CNR SCITEC, Italy
Sofia Pescarin	CNR ISPC, Italy
Irina Crina Anca Sandu	MUNCH Museum, Norway
Saptarshi Neil Sinha	Fraunhofer IGD, Germany
Sophia Sotiropoulou	FORTH, Greece
Panos Trahanias	FORTH, Greece
Giorgio Trumpy	NTNU, Norway
Giovanni Verri	Art Institute of Chicago, USA

PERCEIVE Management Board

Contents

Methods and Workflows for Colour Change Modelling and Reconstruction

Yoko Arteaga[1], Cristiana Barandoni[2], Lucia Burgio[3], David Buti[2], Irina-Mihaela Ciortan[1(✉)], Brenda Doherty[4], Jon Yngve Hardeberg[1], Catlin Langford[3], Donata Magrini[2], Letizia Monico[4], Catarina Pinto[4], Irina Crina Anca Sandu[5], Faïçal Selka[6], Saptarshi Neil Sinha[7], Panagiotis Siozos[8], Sophia Sotiropoulou[8,9], and Giorgio Trumpy[1]

[1] ColourLab, Norwegian University of Science and Technology, Gjøvik, Norway
{yoko.arteaga,irina-mihaela.ciortan,jon.hardeberg,giorgio.trumpy}@ntnu.no
[2] National Research Council, Institute of Heritage Science (ISPC), Florence, Italy
{cristianabarandoni,david.buti,donata.magrini}@cnr.it
[3] Victoria and Albert Museum, London, UK
l.burgio@vam.ac.uk
[4] National Research Council, Institute of Chemical Sciences and Technologies (SCITEC), Perugia, Italy
{brenda.doherty,letizia.monico}@cnr.it, catarina.pinto@scitec.cnr.it
[5] Munch Museum, Oslo, Norway
irina.sandu@munchmuseet.no
[6] IMKI, Strasbourg, France
f.selka@imki.com
[7] Fraunhofer (IGD) Institute for Computer Graphics Research, Darmstadt, Hesse, Germany
saptarshi.neil.sinha@igd.fraunhofer.de
[8] Institute of Electronic Structure and Laser, Foundation for Research and Technology-Hellas (FORTH), Heraklion, Crete, Greece
{psiozos,sophiaso}@iesl.forth.gr
[9] School of Applied Arts and Sustainable Design, Hellenic Open University, Patras, Greece
sophiaso@eap.gr

Abstract. This chapter presents end-to-end methodologies and workflows for modelling, reconstructing, and communicating colour change in cultural heritage across four scenarios: classical sculpture, paintings, historical textiles, and early colour photographs/films. We couple experimental mock-ups, created with chemically appropriate pigments/dyes, with computational imaging and appearance modelling to realistically capture the evolution of the aging process. For sculpture, we propose an atlas-based segmentation and semantic shading pipeline that transfers measured appearance from mock-ups to 3D models, enabling virtual reconstructions of lost polychromy. For paintings, we introduce a dual pathway to predict and visualize light-induced discolouration: a photochemical model linked to radiant exposure and a data-driven spectral unmixing approach; both are demonstrated on Edvard Munch's *The Scream (ca. 1910)*, and further supported by corrections of archival

I. M. Ciortan et al. (eds.), Chromatic Visions, LNCS 16450, pp. 1–56, 2026.
https://doi.org/10.1007/978-3-032-17700-1_1

colour film records. For textiles, we implement appearance-aware 3D restoration and fading simulation, integrating spectral unmixing with multi-temporal photographic documentation to visualize past states and forecast future fading. Finally, for early colour photographs, we present high-magnification multispectral imaging for the recovery of colour in faded autochromes, neural methods for repairing the bleeding of the green dye in some autochromes, and lastly, a refined digital decoding strategy for Kodacolor lenticular film. Together, these methods bridge analytical evidence, appearance capture, and perceptual visualization, providing reproducible, decision-support tools for conservation, display, and public engagement.

Keywords: Photodegradation · Spectral unmixing · Appearance modelling

1 Introduction

Collections of cultural heritage assets that incorporate colour are a priority for museums. Due to their high fragility, the complexity of their study and the significance of their proper communication to future generations, special attention and shared methods are required for curating, preserving, and exhibiting them.

The PERCEIVE project (Perceptive Enhanced Realities of Colored Collections through AI and Virtual Experiences), focuses on coloured collections and aims to empower researchers and museum professionals to achieve their objectives regarding the preservation, authenticity and perception of colour. Colour is one of the most fragile yet meaningful features of cultural objects and heritage collections.

A major objective is to deepen and enrich our knowledge of the original conception of coloured objects and collections. In most cases, depending on the artwork's typology and current preservation state, the colour has changed, faded, or become discoloured compared to the original. Therefore, it is crucial to investigate and identify the original colours, ascertaining not only the chemical composition and provenance of the constituent materials (pigments and binders), but also the original appearance of the coloured surfaces.

Communicating research and interpretation, and the resulting knowledge and understanding, to present and future generations is an equal priority. This knowledge informs debates about authenticity, particularly with regard to reconstructions. The notion of what constitutes the "true" or "faithful" appearance of an object is informed by both material evidence and cultural interpretation. Ultimately, this knowledge enhances perception and interpretation, broadening accessibility to these cultural heritage collections and strengthening public engagement with them, thereby enabling personal and social interaction. However, scientific data is not immediately understandable, interpretable and exploitable for creating pathways that allow us to connect the colour preserved today to its original appearance.

PERCEIVE has explored state-of-the-art technologies to develop digital solutions, tools and services that empower professionals to address their objectives related to the challenges of colour preservation. These challenges include the fragility, change, loss or misrepresentation of colour in artworks and historical objects. The developed tools are presented in Chap. 2 of this book and are based on methods for exploring scientific data to model colour change. The aim is to simulate the effect of deteriorating factors over time in order to either track changes back and reconstruct the original appearance, or predict further change or loss, with the ultimate goal of preventing or slowing down any ongoing degradation process. As well as being used directly, simulations and colour change models allow us to visualise changes and use this visual material to open dialogue and support decision-making among scholars, researchers, and stakeholders.

This chapter is divided into sections, each dedicated to one of the PERCEIVE scenarios, which refer to specific colour change challenges. Colour change can manifest in several ways, such as colour loss, fading and alteration, and the different cases depend on the typology of the object as well as various deteriorating factors. Light has the most significant impact on colouring materials, but this is often exacerbated by the synergistic influence of other weathering parameters or conditions. This dictates a different approach to reconstructing the original appearance in each case. When the colour has barely been preserved on the artwork, identifying the original materials guides the fabrication of mock-ups representing the original painting techniques. The analytical data acquired from the mock-ups or the preserved colours in the case studies were used to develop data-driven models.

2 Polychromy of Classical Sculpture

The persistent myth of a monochromatic aesthetic in classical antiquity has been systematically challenged by modern research.

The study of lost polychromy extends beyond mere aesthetic interest, representing a critical scientific field that focuses on the study of primary and secondary sources, drawing on and combining approaches and methodologies from different disciplines with the common goal of understanding and scientifically re-tracking the effects of material deterioration in order to recover the original visual language of ancient art.

This research combines advanced scientific analysis with rigorous archaeological investigation. Beyond simply identifying pigment residues, this interdisciplinary approach enables a deeper understanding of how colours were applied to statues and architecture. This is essential for comprehending and subsequently communicating meaning, status, and emotion. Ultimately, this research reveals a more complex and dynamic representation of ancient civilizations than was previously thought possible.

The digital reconstruction of lost polychromy on ancient artefacts represents a significant subfield of virtual archaeology and cultural heritage preservation. Early attempts at digitally reconstructing lost polychromy on ancient artefacts were often criticised for lacking authenticity and appearing unrealistic. These models frequently relied on colours that were spread out evenly, resulting in a flat approach that failed to capture the subtle visual appearance properties of real-world materials like gloss and roughness.

Our current approach aims to overcome the limitations of standard reconstruction methods by focusing on achieving more authentic colour representations.

2.1 Material Appearance of Lost Polychromy

Original Colours. As part of the PERCEIVE project, a selection of statues from the National Archaeological Museum of Naples (MANN) were chosen for a case study. The museum's extensive collection provided a significant number of examples, creating a rich iconographic context for comparison with related frescoes and mosaics. This comparative analysis, supported by scientific research and historical documentation, formed the basis for reconstructing the original conception of these artefacts in colour (with polychromy).

Two analytical campaigns were conducted at the MANN on selected statues, following a standardized protocol detailed in previous works [17, 18].

2.2 Reconstruction Approach

The workflow for digitally reconstructing lost polychromy on classical sculptures adopted by PERCEIVE combines an experimental and computational approach based on five steps (Fig. 1):

1. Identification of polychromy traces
2. Experimental fabrication of mock-ups
3. Capturing measurements and digital modelling of mock-ups
4. Atlas-based segmentation of a sculpture modelled digitally
5. Application of mock-up models as shaders

Fabrication of Mock-Ups. Although highly fragmented, some residual polychromy remains visible on the statues of the selected iconography, i.e. Venus, with particular interest in the Anadyomene type. The results indicate a wide colour palette used for the hair and drapery. To replicate appearance, a series of mock-ups were created using the identified materials, guided by ancient historical sources like Pliny the Elder (Naturalis Historia), Vitruvius (De Architectura), and Theophrastus (De Lapidibus) who provided essential descriptions of ancient painting techniques and pigment recipes [23, 31, 34].

The discovered colour palette consisted of a few key pigments, used either individually or in binary and ternary mixtures. Iron-based yellow and red ochres were identified as the primary pigments for the hair. Egyptian blue was used to create blue shades, sometimes mixed with another pigment or applied as a preparatory layer with a white pigment. In the case of the Anadyomene Venus

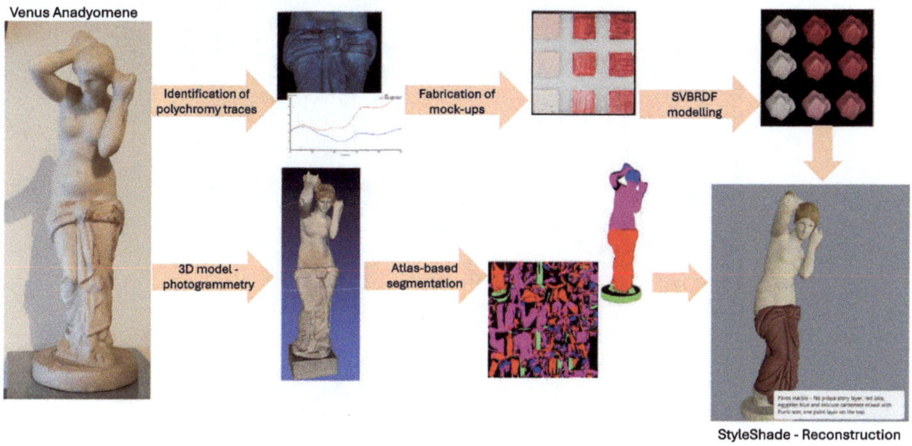

Fig. 1. Reconstruction workflow based on five steps: identification of polychromy traces, fabrication of mock-ups and SVBRDF modelling, 3D model photogrammetry and Atlas-based segmentation. (Color figure online)

from the Iseum of Pompeii (inv. 6298), the research allowed to discover on her the drapery Egyptian blue combined with a madder lake, likely to achieve a pinkish hue. Green earth was also identified on the Lovatelli Venus (inv. 109608).

To recreate the appearance of original polychromy, two replicas of the same set of paint mock-ups were fabricated on Pentelic and Paros marble (Fig. 2). Selecting a substrate similar to that in the original case study aimed to reproduce the texture, translucency, and overall impact of the marble on the final colour. The replicas were designed to simulate the complex interplay of light between the marble surface and the overlying paint layers, as well as the effect of the natural veins in the marble. To this end, a variety of pigment combinations and stratigraphies were created, adhering as closely as possible to the descriptions found in ancient sources.

The paints were applied to 36 marble cubes, each measuring $5 \times 5 \times 3$ cm. The surface of each cube was lightly sanded to create an optimal texture for pigment application. Each surface was then divided into nine distinct squares, each measuring approximately 1.25×1.25 cm. This allowed different combinations of pigments to be applied to separate and limited areas (Fig. 3).

The painting layers were applied using either a ready-to-use Punic wax (from Kremer) or egg white as binding media. The selected pigment was mixed with egg white to create a priming layer.

The pigments chosen for the replicas represent the colours commonly used in the polychromy of classical Roman sculptures. These colours include red ochre, yellow ochre, green earth, Egyptian blue, madder lake, cinnabar, lead white, and calcium carbonate. Lead white was used exclusively for the preparatory layer, while calcium carbonate was also added to various paint mixtures.

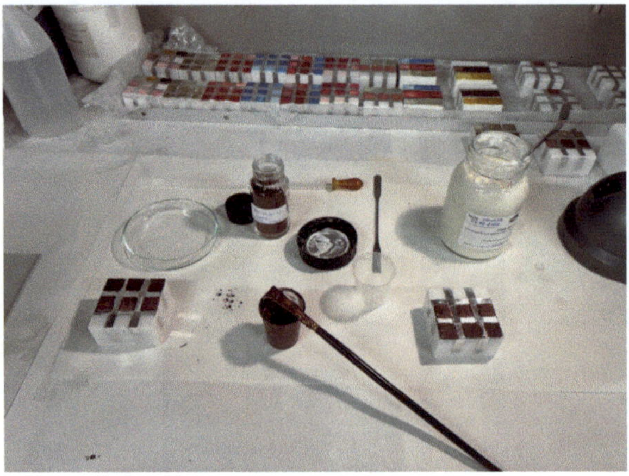

Fig. 2. Fabrication process of the mock-ups.

The marble cubes were painted following a structured approach simulating different application methods:

- **Ground Layer Options**: Four different preparatory layers were used to create the replicas: (a) no ground layer; (b) a calcite (calcium carbonate) ground layer; (c) a lead white ground layer and (d) a mixture containing a small amount of highly dispersed Egyptian blue in calcite.
- **Layering:** Pure pigments were dispersed in Punic wax and applied in single, double, or triple superimposed layers on the ground layer options.
- **Lightness variations:** In addition to pure pigments, mixtures of pure pigments with varying ratios of calcium carbonate were also applied.
- **Pigment Combinations:** Various pigment combinations of pigments were also tested, including madder lake mixed with calcium carbonate, green earth mixed with Egyptian blue, and red ochre layered over yellow ochre.

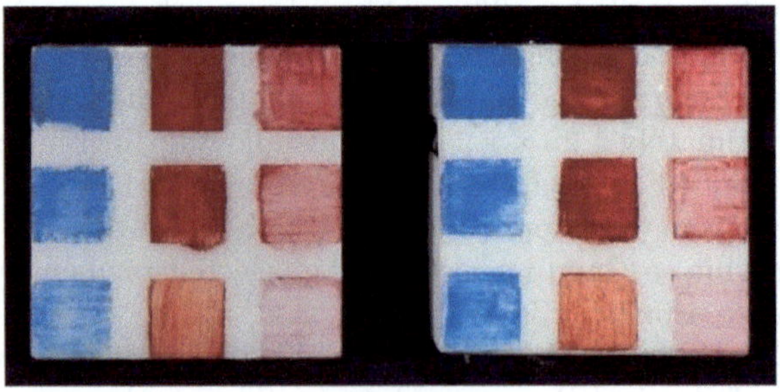

Fig. 3. Example of some mock-ups. The dimensions of the marble cube are 5×5 cm, and each patch has a dimension of approx. 1.25×1.25 cm. (Color figure online)

Material Appearance Modelling. To capture the appearance of the mock-ups' materials, we measure and model the spatially varying bidirectional reflectance distribution function (SVBRDF). The mock-ups were measured using an X-Rite MA-T12 multi-angle spectrophotometer. The MA-T12 instrument measures at six illumination sources and two pick-up points. Surface reflectance is measured in the range of 400–700 nm at 10 nm intervals. The illumination source is a polychromatic white LED with enhancement in the blue region of the spectrum, and its spot size is 9 mm by 12 mm. The MA-T12 also measures six texture images at an angle of incidence of 15° (Fig. 4).

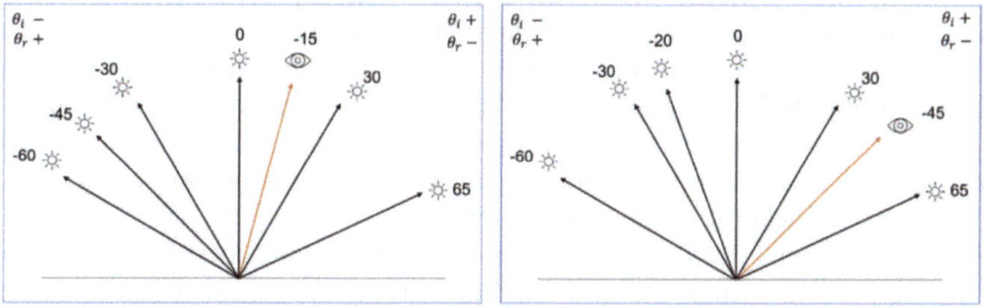

Fig. 4. Schematic of the MA-T12 illumination and acquisition angles. Taken from [1].

The twelve spectra together and the six texture images, are used in the X-Rite Pantora software [36] to obtain the material appearance models. Pantora performs a non-linear fitting to model the SVBRDF to the data and obtains the model parameters in the form of material maps. These are a three-channel diffuse colour map, a normal map, and a specular colour map, and a greyscale roughness map. Pantora also generates a Material Exchange Format (AxF) file [21] which can be used to render the appearance of the material (Fig. 5).

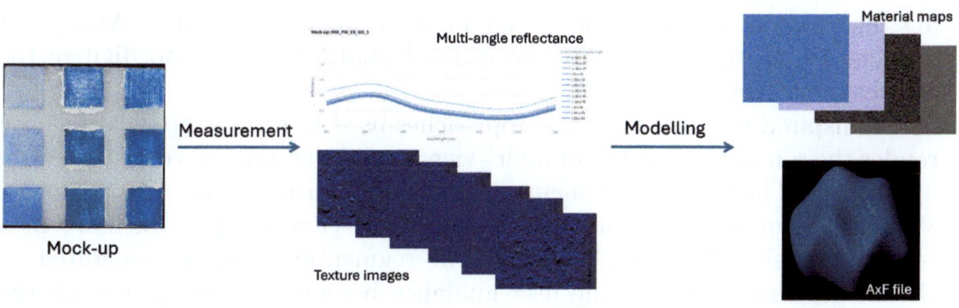

Fig. 5. Pipeline for SVBRDF modelling. The multi-angle reflectance and texture images of each mock-up are measured to obtain the SVBRDF in the form of an AxF file and a set of material maps.

The SVBRDF provides a comprehensive representation of how a material reflects light, accounting for both spatial and angular variations in reflectance. This level of detail is crucial for accurately reproducing the visual properties of real-world materials, particularly those with complex surface features such as gloss, texture, and roughness. By capturing both spectral and spatial information, the MA-T12 enables the creation of digital material models that closely match the appearance of the physical samples under a wide range of lighting conditions.

Once the data is processed in Pantora, the resulting material maps provide a versatile digital representation of the surface. The diffuse colour map encodes the material's base colour, while the normal map captures fine-scale surface geometry that affects how light interacts with the surface.

The specular colour map and roughness map together define the glossiness and specular highlight behaviour, enabling realistic rendering in virtual environments. The AxF file format is a proprietary format and is compatible with a range of rendering engines and design software, facilitating the integration of these material models into digital workflows.

The measurement data, as well as material maps and AxF files, have been uploaded as part of a publicly available Polychromy Appearance database with over 200 mock-up materials. This database has been implemented on Streamlit [29] and it allows the user to search for the mock-up they are interested in, in an interactive way. With drop-down menus, the user can narrow their search by querying different categories such as type of marble, type of binder, pigment, ground layer and number of paint layers. The user can view and download this data to generate their own shaders based on the material maps or directly use the AxF file for their desired application.

2.3 Atlas-Based Segmentation and Semantic Shading

The segmentation (see Fig. 6) and semantic shading of the different segments of the 3D model involves the following steps:

- 3D digitization of the statue using photogrammetry or other AI/non-AI approaches. In this case study, we utilized photogrammetry to digitize the Anadyomene Venus.
- Then, inspired by path planning approaches used in robotic applications, we render the minimum number of multi-view images required to cover the entire 3D surface. The entire implementation was done in Blender.
- We select a few images from the rendered multi-view images, which are then segmented using Segment Anything. The segmentation masks generated by the AI are then refined using user guidance before being tracked to all the segments using an AI-based zero-shot tracker.
- The segmented multi-view images are then projected back to generate a 2D parametrised segmentation map over the 3D surface. We name this process 'segmentation atlas'.

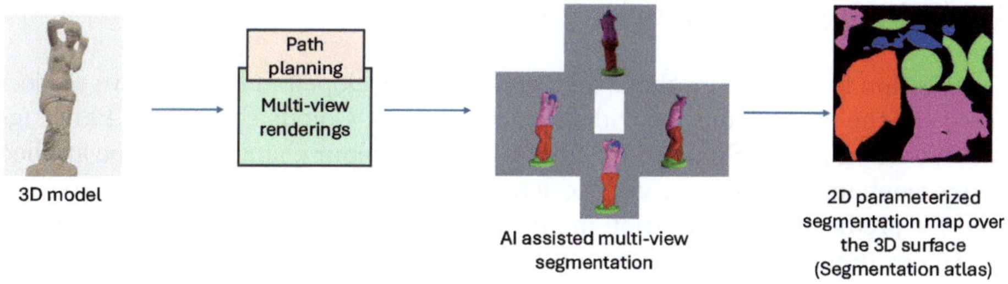

Fig. 6. Pipeline for generation 2D parametrised segmentation map from the 3D model.

This 2D representation of the 3D surface (segmentation atlas) is used to apply the different materials captured by the spectrophotometer as previously explained. We use a modified Cook-Torrance shading model to apply the different material-maps generated by the material capture process. This allows us to apply different materials to the different segments and perform various simulations of the restoration process with each segment. Figure 7 shows a particular example using our approach:

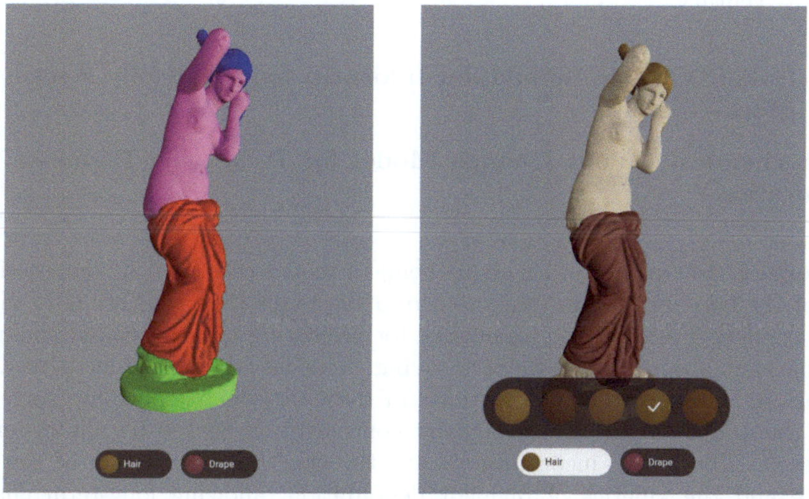

Fig. 7. Segmentation and Reconstruction Hypothesis. On the left the segmentation of the statue using non-representative colours to define distinct segments. On the right one possible reconstruction hypothesis, utilizing material modelled from the paint mock-ups, applied to the segments.

3 Paintings and Works on Paper

This section focuses on methods and workflows developed in the Perceive project based on the case studies provided for Scenario 2: Colour changes in Paintings and Works on paper. Simulation of light-induced colour change had to be applied on pigment mock-ups first, via accelerated aging. Then, one of the approaches used for predicting discoloration in paintings, models changes of the mockups, in the colour domain as linear functions, and then transfer them to a real painting. The other approach models the changes in the spectral domain, using hyperspectral images of the painting as input, where spectral unmixing is used to correlate the mockups with the pigments distribution in the real painting. Beside mock-ups, both methods have been successfully tested on *The Scream (ca. 1910)* by Edward Munch.

Another complementary computational method is presented for studying the temporal evolution of *The Scream (ca. 1910)* and its changing appearance due to the effects of light, moisture, and material ageing. This approach focuses on colour reconstruction using four historical colour photographs from 1971, 1989, 1993, and 2003. To address the challenges of film degradation, the method employs stable reference areas of the painting for advanced colour correction, enabling more faithful digital renderings of its appearance at several points in the 20th century.

3.1 Colour Change Simulation Informed by Artificially Aged Mock-Ups

Photochemical Colour Change Model for Predicting Pigments Discolouration

General Description. The photochemical model presented in this section is one of the two approaches integrated into the Light Damage Estimator (LDE) tool. It offers a mechanistic framework for predicting pigment colour change by linking light absorption to the underlying chemical reactions that drive degradation. In this approach, colour shifts in the CIE L*a*b* space are correlated with the gradual reduction of pigment concentration caused by photochemical processes. The model incorporates both the spectral distribution of the light source and the kinetics of pigment degradation, allowing long-term changes to be expressed as a function of radiant exposure [28]. Unlike the statistical model, which relies on empirical fitting of observed data, the photochemical model offers a process-based interpretation that connects measurable colour alterations directly to the physicochemical mechanisms of light-induced pigment transformation.

Model Workflow. The colour change methodology is organized into two complementary workflows (Fig. 8). The first workflow focuses on the experimental determination of model parameters using artificially aged pigment mock-ups,

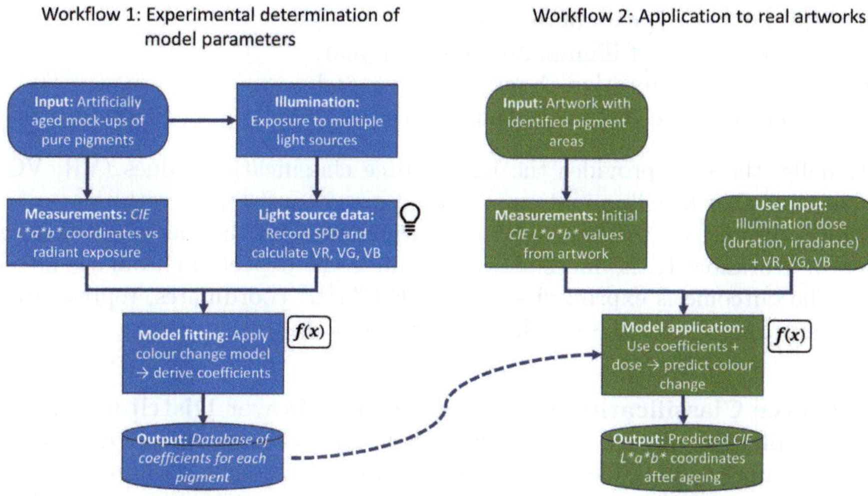

Fig. 8. The two workflows of the photochemical colour change methodology.

while the second applies these parameters to real artworks to estimate colour alterations after prolonged illumination.

Workflow 1: Experimental Determination of Model Parameters. The initial workflow involves the systematic collection of experimental data from artificially aged mock-ups prepared with pure pigments. Each pigment is exposed to controlled artificial ageing using several light sources, each characterized by distinct spectral features. For every light source, the CIE L*a*b* chromatic coordinates are measured at successive ageing intervals and plotted against the corresponding radiant exposure values. In parallel, the spectral power distribution (SPD) of each light source is recorded, and the associated light source classification values (VR, VG, VB) are calculated [28]. These datasets provide the foundation for fitting the colour change model. By applying the model to the CIE L*a*b* coordinates of the artificially aged mock-ups, linear regression yields the coefficients a and b of Equation (2). These coefficients quantitatively describe the colour evolution of each pigment under specific lighting conditions. Once derived, they are stored in a database and subsequently used for predictive simulations on real artworks.

Workflow 2: Application to Real Artworks. The second workflow applies the model directly to cultural heritage objects. The process begins with the identification of specific-coloured regions in the artwork that correspond to pigments previously characterized in the laboratory. From these regions, the initial CIE L*a*b* chromatic coordinates are extracted. To simulate long-term colour changes, the user specifies the illumination dose that the artwork will experience. This dose is calculated based on:

- the total number of years of display,
- the average number of illuminated days per year,
- the daily exposure duration (hours per day), and
- the irradiance of the chosen lighting source.

Additionally, the user provides the light source classification values (VR, VG, VB), ensuring that the illumination is characterized consistently with the experimental conditions. By integrating these parameters with the model coefficients derived in Workflow 1, the methodology predicts the expected chromatic alterations. The outcome is expressed as new CIE L*a*b* coordinates, representing the pigment colour after the simulated ageing period.

Light Source Classification Based on Spectral Power Distribution. We adopt a simplified approach for classifying light sources according to their spectral characteristics. This method is designed to be accessible to users without specialist expertise. The classification scheme categorizes light sources based on the distribution of their emission across spectral ranges corresponding to the three primary colours of light as perceived by the human eye within the visible spectrum.

Table 1. Spectral ranges selected for the calculation of a light source percentage of power distribution.

Classification	Low (nm)	High (nm)
Red spectral range (R)	380	500
Green spectral range (G)	500	600
Blue spectral range (B)	600	780

This approach is consistent with museum lighting practices, where emissions outside the visible range are typically filtered out. As a result, each light source can be described using only three values, which represent the percentage of power emitted in the red, green, and blue spectral regions of the visible range (Table 1). For each light source, the values VR, VG, and VB correspond to the fraction of emission within the respective spectral range relative to the total emission across the 380–780 nm interval:

$$V_{R,G,B} = \frac{\text{Red, Green, Blue spectral range}}{\text{Total emission in visible (380–780 nm)}} \times 100\% \qquad (1)$$

The model presumes that colour changes induced by light are determined mainly by the intensity of radiation within the spectral band to which a pigment is most responsive (VR, VG, or VB), rather than by the total irradiance across the visible spectrum. Said otherwise, when a pigment is exposed to light concentrated in its most sensitive wavelength region, the extent of colour alteration will be similar to that observed under full-spectrum illumination, provided the effective portion of irradiance remains equivalent. Based on this, photodegradation arises specifically from the segment of the spectrum that the pigment absorbs.

Photochemical Colour Change Model. The model implemented in the Light Damage Estimator (LDE) tool assumes that colour changes in a material, expressed as shifts in the L*, a*, and b* values (which correspond to variations in the visible diffuse reflectance spectra), are primarily driven by photochemical reactions affecting the pigment. These reactions alter the pigment's chemical composition, leading to a reduction in its original concentration. The model therefore correlates colour change directly with the decrease in pigment concentration. It is important to note, however, that other factors—such as loss of surface gloss or yellowing of the binding medium—may also influence colourimetric measurements. While these effects are not accounted for in the model, they should be considered when interpreting results. When exposed to light, the material absorbs radiation, triggering photochemical reactions that gradually reduce pigment concentration. At low levels of light absorption (zero-order kinetics), the rate of this reduction is proportional to the absorbed energy. As pigment concentration decreases, the fraction of incident radiation interacting with pigment molecules also changes, influencing the rate of degradation.

Assumption 1. Photochemical reactions in pigments are generally slow, producing only minor changes in pigment concentration during typical exposure times. This assumption is valid for gradual degradation processes, where pigment concentration remains nearly constant over the observation period. It also provides a useful metric for assessing the sensitivity of colour or reflectance shifts to small pigment losses. By tracking reflectance or colour values over time, even subtle degradation can be detected and quantified.

Assumption 2. Any colorimetric change C, whether expressed in the CIE Lab* colour space, as ΔE_{1976}, or as diffuse reflectance variations, is assumed to be proportional to the pigment concentration n in the paint. From this assumption, the following expression is obtained:

$$C = -\frac{1}{A} \ln\left(\frac{\varepsilon n_0}{\Phi}\right) + \frac{1}{A} \ln(E_0) = a + b \ln(E_0) \quad (2)$$

In this formulation, the quantum yield Φ represents the efficiency of the photochemical process, defined as the number of molecules degraded per photon absorbed. Although Φ could, in principle, be experimentally determined through chemical or spectroscopic analysis, in the present model it functions as a proportionality factor embedded within the slope of the linear relationship between colour change and the logarithm of radiant exposure (see Eq. 2). Consequently, both Φ and the molar absorptivity ε are not measured directly but are incorporated implicitly through empirical validation, thereby supporting the model's applicability to light-induced colour change data derived from mock-up experiments. As currently formulated, the model is optimized for single pigments or simple binary mixtures in which one pigment exerts dominant control over the observed colour. Extending the model to more complex systems, such as multi-pigment mixtures, would require accounting for the distinct spectral sensitivities, degradation kinetics, and potential interactions among different pigments. This includes modeling how each pigment contributes to overall reflectance and how their individual photodegradation pathways evolve—either concurrently or competitively—over time. The model also assumes a constant oil-to-pigment ratio and attributes colour changes primarily to pigment degradation. In practice, however, variations in binder-to-pigment ratios can significantly affect optical properties such as surface gloss, scattering, absorption, haze, and translucency. Under these conditions, degradation of the binder may have an equally or even more pronounced impact on colour change than pigment alteration alone.

Calculating the Model's Coefficients for a Pigment. Light-induced colour change in the paint is determined not by the total irradiance over the entire emission spectrum, but by the irradiance in a specific spectral region that corresponds to the part of the spectrum that the pigment absorbs, i.e. the spectral region that is most sensitive. In this context, it is not necessary to consider the total irradiance that illuminates the material. Instead, only the fraction corresponding to the spectral component that affects the pigment should be considered. To determine the spectral component affecting the pigment, we plotted the values of ΔL^*, Δa^*, and Δb^* as a function of each of the spectral components R, G, B of the radiant exposure for each light source. The three spectral components of radiant exposure were calculated using the following equations:

$$H_{B,G,R} = \frac{V_{B,G,R}}{100} H_{\text{Total}} \tag{3}$$

where H_{total} is the total radiant exposure, V_B, V_G and V_R the percentage of emission in the specific spectral range derived from Eq. 1, and H_B, H_G, H_R the blue, green and red radiant exposures, respectively. To quantify these observations, the colour-change model described in this chapter was applied to establish predictive relationships for colour evolution. Following Eq. 2, the natural logarithm of both the total radiant exposure and the spectral component-specific radiant exposures was calculated. The resulting ΔL^*, Δa^* and Δb^* values obtained under different light sources were plotted against the logarithm of radiant exposure for Prussian blue–lead white paint mock-ups (Fig. 9). In all cases, the colourimetric shifts exhibited a pronounced linear dependence on radiant exposure when represented on a logarithmic scale, thereby supporting the validity of the proposed model. The fitting-derived coefficients a and b for each colour coordinate are subsequently stored and employed to predict colour changes in artworks.

Application of the Colour-Change Model to Artworks. The calculation of colour change in artworks is based on the user-provided input of radiant exposure (light dose) together with the selection of the pigment of interest. These two parameters form the starting point for the prediction process. Using Eq. 2,

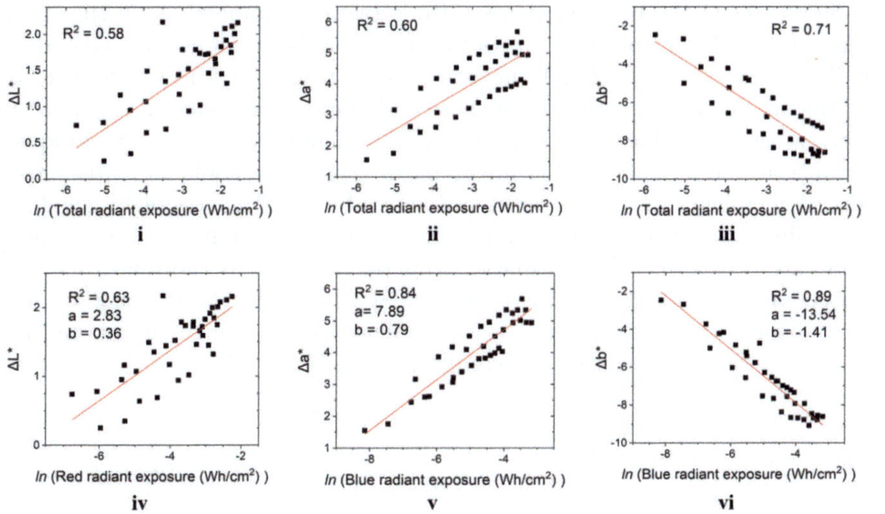

Fig. 9. Plots of ΔL^*, Δa^* and Δb^* values obtained with the colour-change model for Prussian blue–lead white oil paints, shown as a function of total radiant exposure (i–iii) and of the spectral radiant exposure components (iv–vi). The coefficients a and b are derived from the linear fits and subsequently stored for use in colour-change prediction. (Color figure online)

the model determines the expected colour shifts, expressed as ΔL^*, Δa^* and Δb^*, for the chosen pigment under the specified illumination conditions. To translate these shifts into perceptible changes, the calculated ΔL^*, Δa^* and Δb^* values are added to the current CIE Lab* coordinates that describe the pigment's colour within the artwork. This operation results in a new set of colour coordinates that represent the predicted appearance of the pigment after exposure to the given light dose. In practice, this approach allows users to simulate the progressive alteration of colour in a specific pigment region of an artwork, thereby providing a quantitative basis for visualising and anticipating the impact of light-induced degradation over time.

Data-Driven Pigment Photo-Degradation Model in the Hyperspectral Domain. In the context of pigment degradation modelling, previous knowledge on the material palette is very meaningful as it facilitates the creation of mock-ups, where swatches of fresh pigments are created to resemble the chemical composition of the artwork, and then artificially aged in a weathering chamber where environmental parameters such as light and relative humidity can be controlled. In this colour change simulation approach, we combine artificially aged mock-ups with a hyperspectral image of the artwork under investigation. More precisely, as shown in the diagram in Fig. 10, we perform spectral linear unmixing in the hyperspectral image, using least-squares minimization and assuming known spectral signatures of the pure pigments (also known as endmembers in the spectral imaging field). After unmixing, we obtain the abundance maps for each endmember. Then, we can reconstruct the painting by re-mixing the abundance maps with the artificially aged spectra of those endmembers included in the mock-ups, following the same linear mixing model. Nonetheless, the mixing techniques in paintings are commonly of non-linear nature, as there could be superimposed layers or chemical mixture between various materials. To account for the limitations of linear spectral unmixing and mixing models when it comes to paintings, we operate in the absorbance space, $-log(R)$, where R is the reflectance. This is based on the findings of Valero et al. [33] who proved that linear unmixing in the absorbance space, $-log(R)$, is as efficient as non-linear unmixing in the reflectance space for paintings' decomposition.

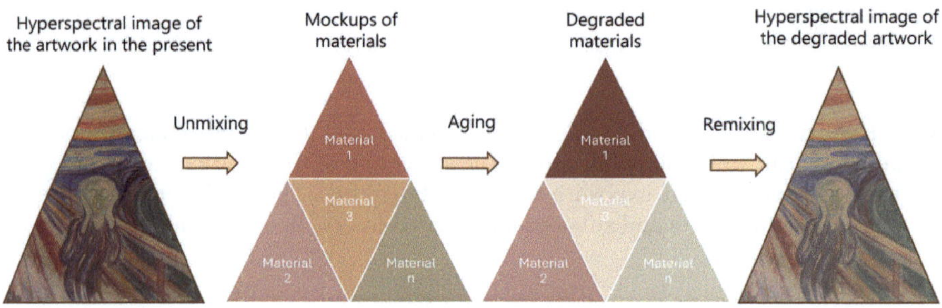

Fig. 10. Our data-driven photodegradation models needs as input a hyperspectral image of the artwork and a set of spectral measurements of the material palette, at various aging levels.

The material palette of *The Scream (ca. 1910)* is known from previous analytical studies and it includes the following pigments: vermilion, cadmium yellow, chrome yellow, viridian, cobalt blue, ultramarine blue, two types of red lake, zinc white, lead white and possibly red lead. To assess light-induced damage and predict future appearance of the artworks, in the context of PERCEIVE several such mock-ups were created by CNR and MOLAB that were later artificially aged. A subset of the palette was chosen for the *The Scream (ca. 1910)* case study according to the transience level of the pigments and so we created mockups of vermilion, chrome yellow, cadmium yellow, zinc white, lead white, and red lead. In addition, the painting was scanned with a hyperspectral imaging system under a MOLAB campaign in 2017.

Thus, to pursue the spectral unmixing task, the spectral signatures of the endmembers in *The Scream (ca. 1910)* have a diverse provenance: some are given by the mock-ups (that only partly cover the painting palette), others are extracted from the spectroscopic measurements on the painting (cardboard substrate, red lakes) and the remaining ones from a privately shared spectral library, with pigments mixed in egg-tempera medium. The reflectance spectra of the endmembers are plotted in Fig. 11. The accelerated aging experiment exposed the mockups to 33.06 $W/cm^2 \cdot hr$ of total light irradiation under a UV-filter Xenon lamp. The spectral measurements were collected at 8 discrete time steps, including t_0. To access unmeasured time steps within the monitored time-span, we interpolate in-between these discrete 8 aging steps. Specifically, an aging polynomial is fitted for each pigment using spline interpolation in the spectral domain, where the dependent variable is the total exposure (in either irradiance or illuminance units) (Fig. 12).

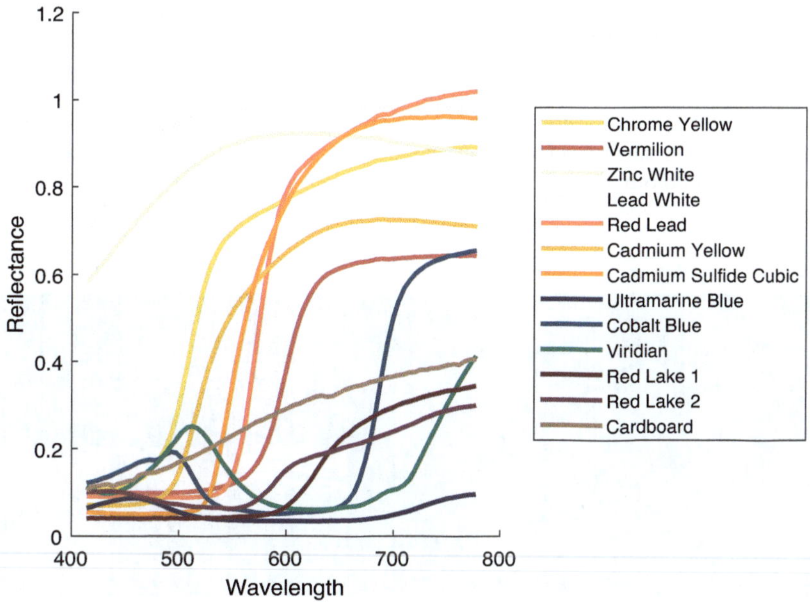

Fig. 11. Endmembers considered for the spectral unmixing of *The Scream*.

The spectral unmixing is carried out in the spectral range from 400 to 850 nm and yields the distribution of pigments over the entire surface of the artwork. The pigment maps (see Fig. 13) show a reasonable distribution and are comparable with the results of analytical techniques. For instance, in the case of the inorganic pigments, such as vermilion the abundance maps are compatible with the XRF maps of the associated chemical elements, mercury (Hg) and chromium (Cr), respectively, and both indicate the presence of the pigments in similar locations, as can be noticed in Fig. 14. Nonetheless, the two abundance maps that represent cadmium yellow in two chemical states (including the cubic form) do not manage to fully capture the cadmium element. This is especially obvious in the depiction of the lake area in the painting, where the yellow ellipse-shaped strokes are erroneously and partially attributed to chrome yellow instead, by the spectral unmixing model. One reason for this could be limited spectral range (visible-near

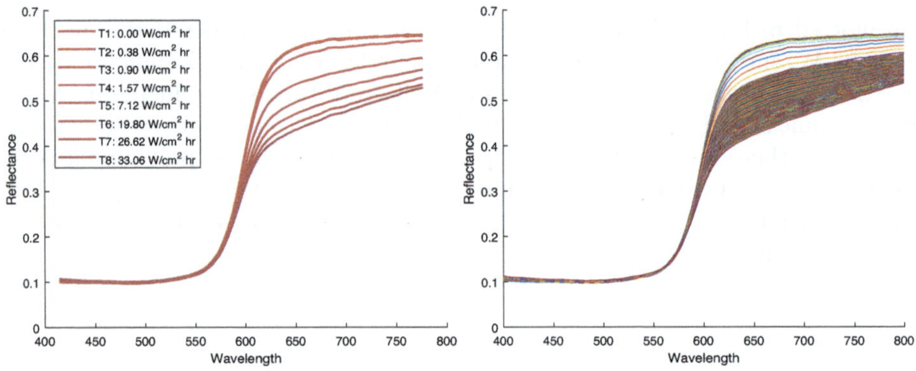

Fig. 12. The aging of vermilion monitored at discrete time steps (left) and the interpolation of intermediate steps, sampled every 0.15 $W/cm^2(right) \cdot hr$. As the light exposure increases, the reflectance decreases in the right tail of the spectrum, meaning that the red colour of vermilion becomes darker and less saturated. (Color figure online)

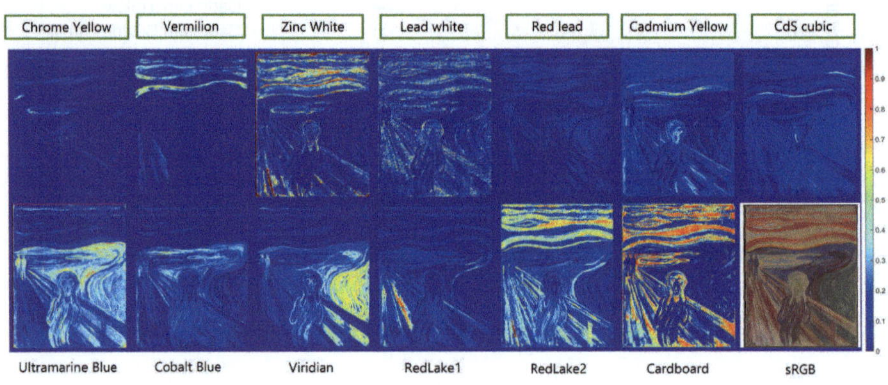

Fig. 13. The pigments maps of *The Scream*, where the red false colour indicates areas with a higher concentration of a material. The photodegradation of the pigments in the top row can be predicted, as aging data was collected for their corresponding mockups. (Color figure online)

infrared) that we operate in, where the yellows have less discriminative features as opposed to the ultraviolet range.

Given the abundance maps and the aging polynomial for some of the pigment, we can now simulate further damage by recombining the pigments maps with the aged spectra. The pigments whose aging behaviour we haven't measured are left unchanged, as in their current state. Figure 15 proposes a degraded version of the painting, equivalent to 34.5 $W/cm^2 \cdot hr$. In the simulation, the brushstrokes in the sky that correspond to vermilion get overall darker, which is a good result, and consistent with our other simulation methods, presented later in this chapter.

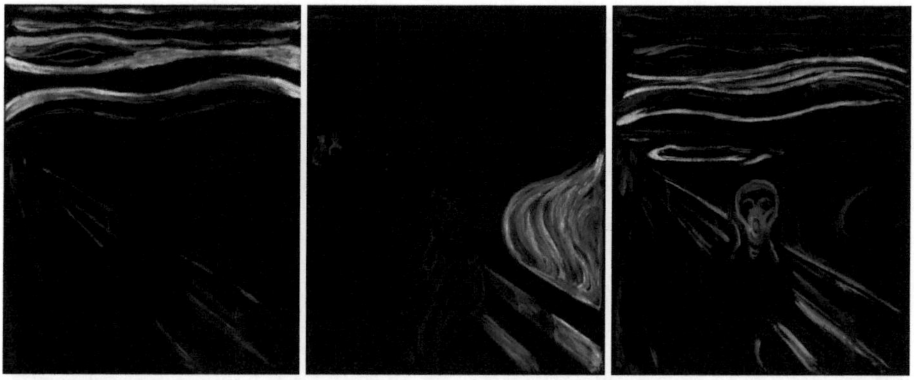

Fig. 14. XRF maps for mercury (Hg), chromium (Cr) and cadmium (Cd) chemical elements, associated with vermilion, viridian and cadmium yellow pigments. (Color figure online)

One limitation of our approach is that not all the pigments in the painting have had aging data collected. Nonetheless, this is a data capture limitation, and not a method limitation, because our method can accommodate additional aging data for all the endmembers.

3.2 Modelling Colour Change Using Film Archival Records as Temporal Reference

Colour Film Modelling. Based on Munch museum's archival records, *The Scream (ca. 1910)* was documented with 4 colour photographs from 1971, 1989, 1993 and 2003, shown in Fig. 16. The challenge is that some of these photographs were themselves subject to degradation, as it is obvious with the one from 1971, where the cyan dye heavily faded. In this approach, we aim to apply a colour correction so that the photographs become more reliable in revealing the past colours of the painting.

In the first step of our approach, we adjust the spectral densities of the cyan, magenta, and yellow dyes so that they match the original concentrations, taking the black level in the film as reference [4]. To this purpose, we need the spectral densities of the photograph, so we captured it with a hyperspectral transmittance imaging setup. The dye adjustment result, displayed in Fig. 17 is a more chromatic balanced film record of the painting, where the magenta colour cast is toned down and the cyan dye is slightly reinjected into the image.

In the second step, we develop a film to painting transformation, where we correct the colours of the film record using the painting as reference. From prior chemical, analytical and accelerated aging investigations, we are aware of which materials and areas in the painting have changed. More precisely, it is known that the vermilion has darkened due to light-exposure, and that cadmium yellow has faded due to humidity. From the XRF analysis, we have the accurate distributions of these pigments in the artwork (see Fig. 14). In addition, the painting

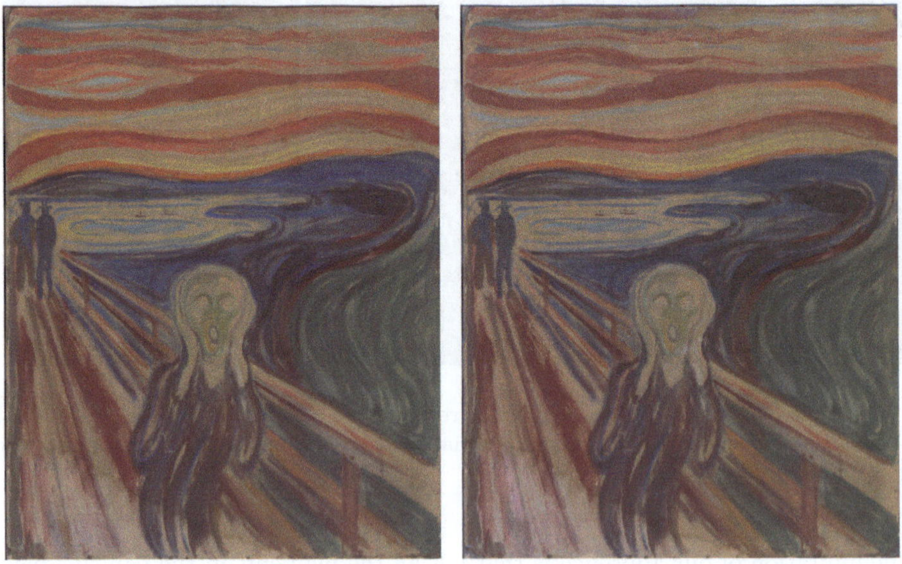

Fig. 15. Current state of The Scream, as scanned in 2017 (left), and the future version, simulated for a total exposure of 33.06 W/cm^2 · hr under a Xenon light source (right). The darkening of vermilion is evident in the brushstrokes in the sky.

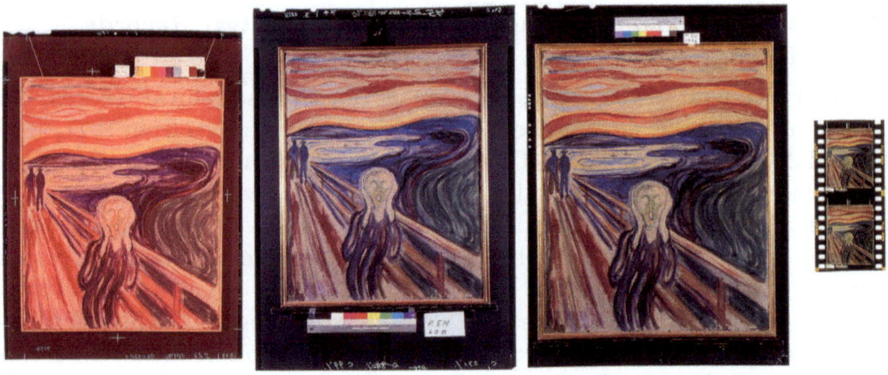

Fig. 16. The analogue photographs of The Scream in 1971 (Kodak Ektachrome), 1989 (Agfa), 1993 (Agfa), and 2003 (Fuji 35 mm).

suffered a water damage in the bottom left corner. We assume that apart from these changes, the rest of the painting is stable and has not changed during the last decades. These stable areas become anchor points to define a transformation from the colours of the slides to the recent colours of the painting last captured in an imaging campaign in 2017, as portrayed in Fig. 18.

Mathematically, we formalize the film-to-painting correction with least-squares regression of polynomial models (first-order, second-order, third-order)

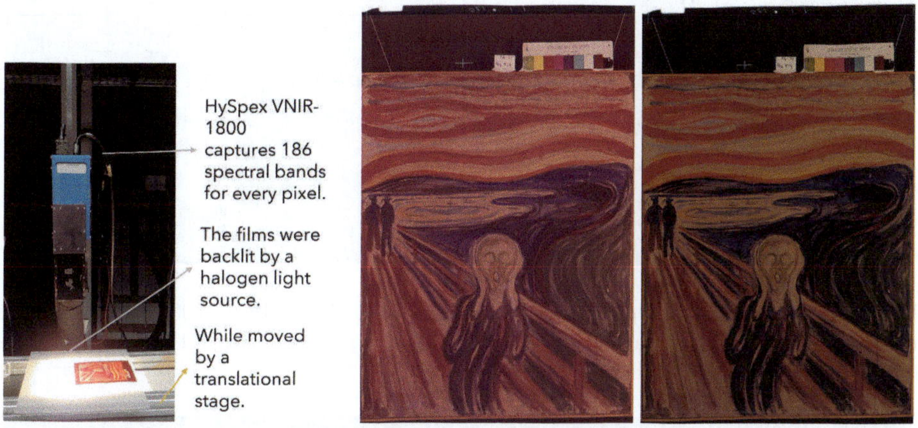

Fig. 17. The hyperspectral transmittance imaging setup (left), the film record from 1971 in its original version (middle) and after pre-processing step with chromogenic dye adjustment (right) that reduces the magenta cast. (Color figure online)

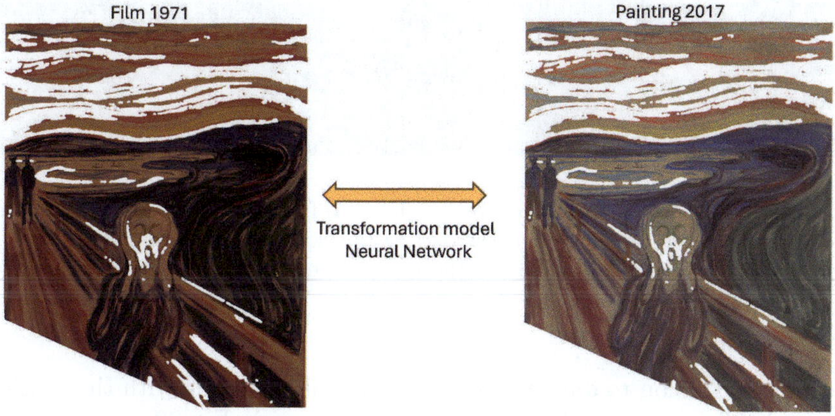

Fig. 18. The stable areas in the painting (not covered by white in the above images) become anchor points for establishing a film-to-paint correspondence.

and neural network fitting with a shallow feed-forward neural network with one hidden layer [7]. We choose the best model according to the mean absolute error given a train-test split of the data, which in our case is the shallow neural network [6]. Finally, we apply the resulting transformation to each film (see Fig. 19), which ultimately enables us to peek into the painting's past with improved colour accuracy.

Generally, one of the challenges of digital restoration approaches resides in the lack of ground truth. A work-around the lack of ground truth is to check the consistency of well-known colour change behaviours of some pigments. We

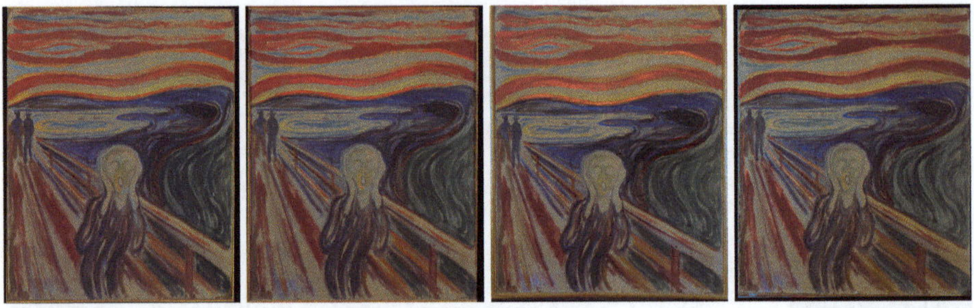

Fig. 19. The corrected film records (in order: 1971, 1989, 1991, 2003) following the neural network transformation based on the anchor points between the recent colours of the painting (2017) and each film.

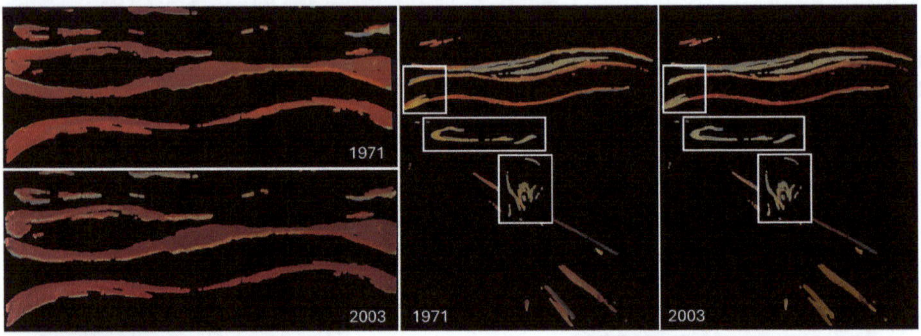

Fig. 20. In the colour corrected versions of the films from 1971 and 2003, we can notice the darkening of the vermilion brushstrokes in the sky, and the fading of the cadmium yellow especially in the lake area, on the face of the character and on the upper left side. (Color figure online)

expect the vermilion to darken and cadmium yellow to fade with the passing of time, and this is perceivable in our results displayed in Fig. 20.

4 Virtual 3D Restoration and Simulation of Fading in light-sensitive Textiles

Heritage textiles are known for containing dyes particularly sensitive to photodegradation. For this reason, their virtual restoration and prediction of further discoloration are of especially high importance towards their documentation and preservation. Digital computation techniques introduce innovative approaches to realistically depict aging processes without intervention on the real objects. In this sub-chapter, we present an approach for modelling and visualizing colour changes in two textiles housed by the Victoria and Albert museum: a 20th-century kimono and 19th-century printed cotton Victorian dress, dyed with natural and synthetic materials, respectively. Our simulations are rendered in the three-dimensional space providing a realistic and interactive environment for scrutinizing the different hypotheses regarding the past and future appearance of the textiles.

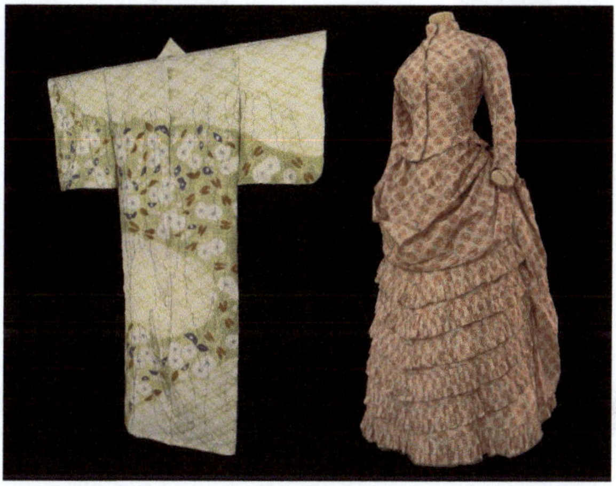

Fig. 21. Snapshots of the digitized 3D models of the kimono and Victorian dress (from the V&A collection) Produced by the V&A Photographic Studio.

Our model combines an *atlas-based material segmentation* with advanced colour restoration techniques to transfer the material appearance from textile mock-ups to the 3D models of the garments [26]. Fading effects are simulated through the accelerated aging of these mock-ups that have the same chemical composition as the historical textiles. The aging process is monitored both colorimetrically and spectrally, while the full appearance attributes are captured using a total appearance capture device (TAC7). We then create a colour-based segmentation map over the 3D surface of the textiles to transfer the appearance of the mock-ups on the segmented regions. The fading simulations are further refined by tracing a correspondence in the spectral domain, between the mock-ups and the current appearance of the textiles. Lastly, we also harness available photographic documentation to speculate on the past colour and texture of the fugitive dyes in the textiles.

We embedded our appearance-aware fading simulations in an interactive application, entitled SimTex (**Sim**ulation of colour changes in **Tex**tiles), where the user can toggle between the current state of the two textiles and the three realistic scenarios for depicting the colour changes (see Figs. 22 and 23), as follows:

1. **Mock-up Appearance**: This mode allows for the application of material maps from both aged and unaged mock-ups to the fugitive colors, facilitating appearance-aware construction.

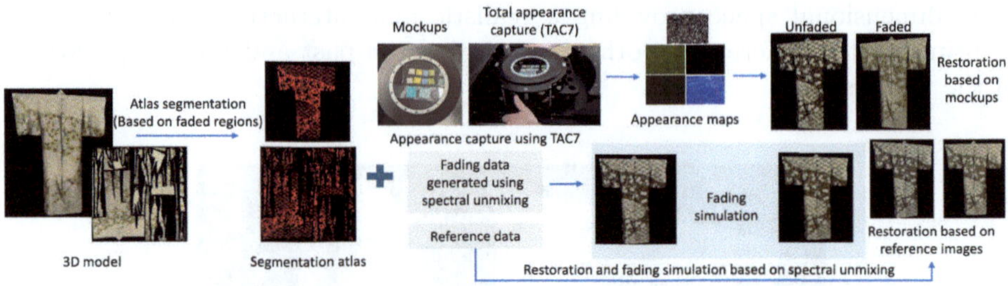

Fig. 22. The segmentation of fugitive colors in the UV-parameterized 3D model is achieved using a 2D parameterized atlas. This segmented atlas enables fading simulations and appearance-aware restoration of fugitive colors, guided by mockups, fading data, and reference images. Image source [26].

2. **Spectral Unmixing**: This mode combines accelerated aging data from the mock-ups with spectral unmixing techniques, revealing the correlation between the fading of the mock-ups and the historical object. It is represented by texture images that illustrate a sequence of fading applied to the segmented fugitive colors.
3. **Photographic documentation**: This mode enables the application of textures from reference documentation images that depict the object in the past, allowing for visualization of potential restoration. The museum's archive contains multiple digital images showcasing details of the kimono. The earliest image is from 2006 (front view), followed by one from 2019 (front view), and a 2024 capture that reveals the inner surface of the inner folds, previously shielded from light exposure. Additionally, there is a 2024 digital image of the inner surface of the Victorian dress.

The interactive apps for the two textiles can be found at the following permanent URLs: https://perceive-tools.igd.fraunhofer.de/kimono/ and https://perceive-tools.igd.fraunhofer.de/dress/.

In the remainder of this sub-chapter, we will elaborate on all the stages in our methodology.

4.1 3D Model Capture and Atlas-Based Segmentation

In this section, we describe our process for digitising the textiles in 3D and conducting automatic segmentation of the fugitive regions.

Photogrammetry Capture for 3D Model Generation: The Victorian dress and the kimono (see Fig. 21) were positioned on a mannequin and a T-bar structure, respectively, within a photographic studio. Continuous halogen lights were employed, remaining fixed in their positions throughout the capture process to ensure consistent lighting. The lighting was kept soft and non-directional

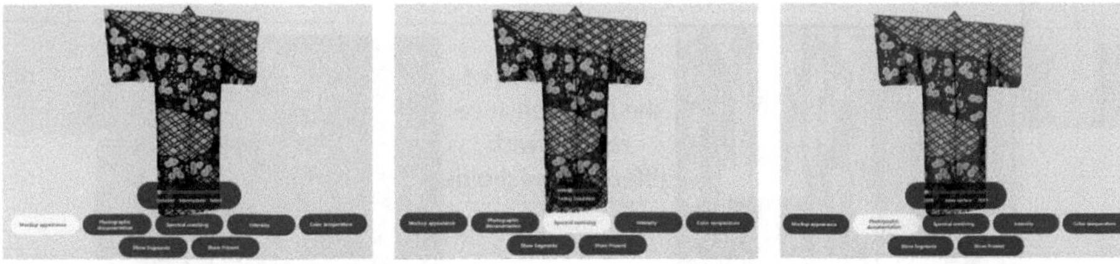

Fig. 23. The figure above illustrates the three visualization modes available in the SimTex interactive visualization tool: Mock-up Appearance (left), where the material appearance from the mock-ups is directly transferred; Spectral Unmixing (middle), which simulates fading based on the current concentrations of the fugitive dyes; and Photographic Documentation (right), which utilizes colour data from various archival photographic records.

to minimize harsh shadows on the objects. A Sony Alpha 7 IV hybrid camera paired with a Sony 50 mm f1.4 lens was utilized for the capture, ensuring over 60% overlap between images in both the x and y directions. The focus was set and maintained consistently throughout the entire capture process. Agisoft Metashape Professional software was used for image processing.

Atlas-Based Segmentation. The 3D modelling of the kimono and Victorian dress using photogrammetry provided us with a texture atlas for both textiles. A texture atlas is a single composite 2D image that packs multiple sub-textures into one image file for efficient rendering. These sub-textures define the UV parameterization of a 3D surface by assigning each mesh vertex (x, y, z) specific UV coordinates (u, v) within the atlas. When the atlas additionally encodes semantic regions, it is referred to as a segmentation atlas.

To create the segmentation atlas, we apply segmentation in the HSV (Hue, Saturation, Value) colour space, with thresholding to identify and isolate the faded regions within the parameterized texture atlas, as exemplified in Fig. 24 for the kimono. For the *i-th* faded region, the thresholds are defined as $H_{min}^{(i)} \leq H_{max}^{(i)}$, $S_{min}^{(i)} \leq S_{max}^{(i)}$, and $V_{min}^{(i)} \leq V_{max}^{(i)}$ for hue, saturation, and value, respectively. The overall condition for segmenting a pixel in the *i-th* faded segment can be expressed as:

$$(H_{\min}^{(i)} \leq H \leq H_{\max}^{(i)}) \wedge (S_{\min}^{(i)} \leq S \leq S_{\max}^{(i)}) \wedge (V_{\min}^{(i)} \leq V \leq V_{\max}^{(i)}) \qquad (4)$$

Finally, we developed a custom shader that utilized the UV-parametrized segmentation atlas to achieve the desired appearance for the various segments of the heritage garments.

Improved Segmentation Based of Structural Features. Segmenting fine-grained features is challenging when colours are similar or structures are very small. In our Victorian-dress example, we therefore relied on structural features

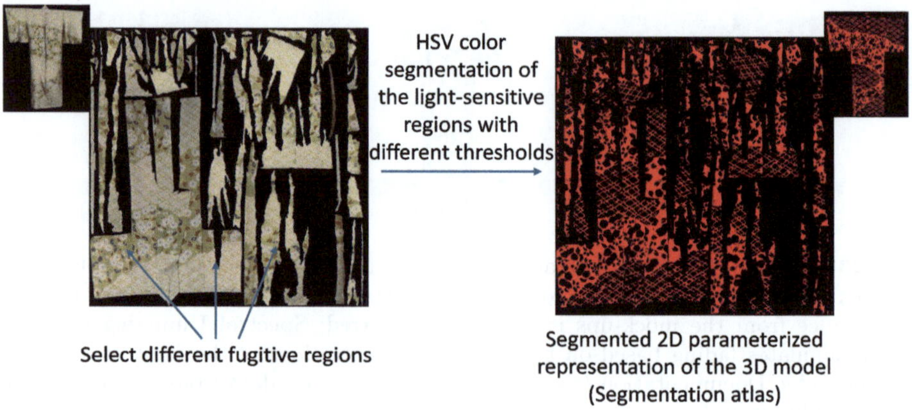

Fig. 24. Segmentation atlas generation through HSV colour segmentation of the faded regions using various thresholds. Image source [26].

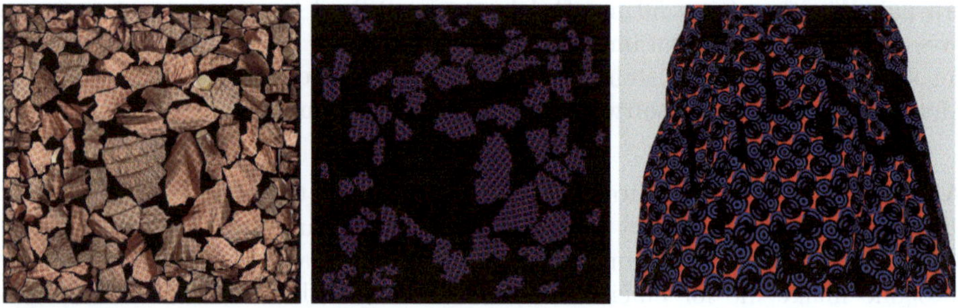

Fig. 25. Segmentation of the Victorian dress using structural features: (left) original atlas, (middle) registered features in the atlas, and (right) resulting partial segmentation when rendered in false-colour for a particular camera view (here blue represents faded purple and red the faded turquoise). (Color figure online)

rather than colour alone. By registering those structural features to a reference atlas, we were able to isolate the faded purple (blue-labeled) and turquoise (red-labeled) regions (see Fig. 25). Although this method yields a partial segmentation of the dress, completing the full segmentation remains future work where we intend to combine colour and structure based segmentation.

4.2 Fabrication of Mockups

The textile mock-ups were prepared as representative models of the historic kimono and dress, designed to address the specific research questions of each case study with regards to their fading and to allow controlled experimentation without risking the original garments. By closely replicating materials, techniques, and visual appearance, the mock-ups provide a platform to study ageing and degradation under defined environmental conditions, particularly light exposure, offering crucial insights for preventive conservation.

The process began with detailed on-site, non-invasive analyses of each textile, including fibre identification, dye composition, and assessment of conservation state. These results guided the selection of historically accurate materials and dyeing techniques. Bibliographic documentation ensured that key variables such as mordanting methods, dye application sequences, and environmental history could be reproduced.

Once completed, the mock-ups were subjected to artificial photoaging under museum-specific lighting conditions. The exposure regimes replicated recorded lux levels, spectra, and durations from actual display periods, making the results directly applicable to the original artefacts. This approach enables prediction of material behaviour, investigation of degradation pathways, and evaluation of preventive conservation strategies in a reproducible, risk-free context.

Case Study: Myriad Green Leaves Kimono

The silk kimono, created in 1992 by the Japanese artist Furusawa Machiko, is part of the Victoria and Albert Museum's textile collection (accession number FE.422-1992). It was crafted using shibori, a traditional Japanese resist-dyeing technique involving precise stitching, gathering, and binding to protect areas of fabric before successive dye applications. The dominant green hue was achieved through a two-step process: initial dyeing with yellow Kariyasu (Miscanthus tinctorius), followed by over-dyeing with indigo (Indigofera tinctoria), as recorded in the artist's notes.

To replicate this textile, silk samples were prepared using artisanal dyeing methods with natural extracts of Kariyasu (yellow), indigo (blue), and a sequential application of both dyes to produce green. Both mordanted and non-mordanted fabrics were prepared to reflect possible historical variations in dyeing practice, as shown in Fig. 26. An undyed white silk control sample was also included for baseline measurements.

For artificial aging, one set of each colour (white, yellow, blue, and green) was stored in darkness as a reference, while the other set underwent 259 h of controlled light exposure, simulating the museum's halogen and LED display conditions at 50 lx. Colour changes were measured periodically using colorimetry and reflectance spectroscopy, enabling the monitoring of fading and the detection of underlying chemical transformations.

Case Study: Victorian Printed Cotton Dress

The second case study focuses on a cotton dress printed by Edmund Potter & Co. in 1887, a leading calico printer of the 19th century (accession number T.2-1926). The design features a turquoise-blue background with motifs inspired by Japanese kamon and Indian chintz. Over time, sections of the fabric exposed to ambient light have faded markedly, while areas protected by folds retain most of their original brightness, as can be noticed in Fig. 27.

Material analysis suggests the use of early synthetic dyes from the triarylmethane family, common in late 19th-century textile production. To reproduce the dress's colour palette, cotton samples were mordanted and hand-dyed (see

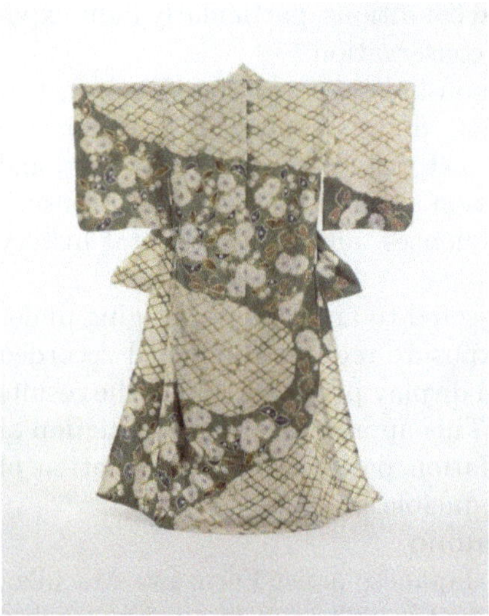

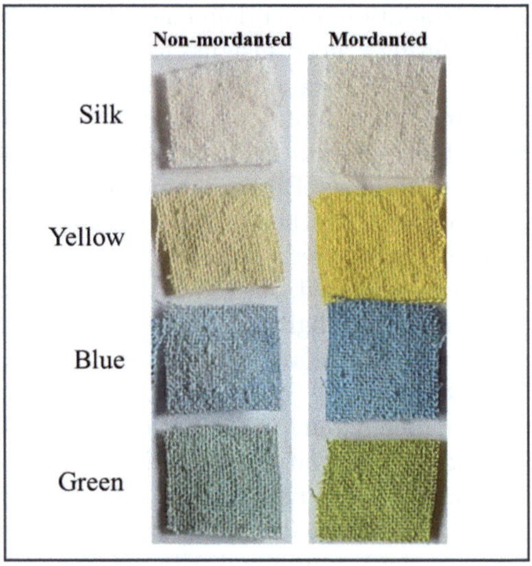

Fig. 26. Left: The *Myriad Green Leaves* silk kimono (Furusawa Machiko, 1992), Victoria and Albert Museum (FE.422-1992). The layered dyeing process using Kariyasu and indigo produces the characteristic green hue. © Victoria and Albert Museum. Right: Silk mock-ups replicating the kimono's dye palette and methods, prepared for artificial aging tests. (Color figure online)

Fig. 27) with diamond green B (Basic Green 4, C.I. 42000), diamond green G (Basic Green 1, C.I. 42040), and yellowish light green SF (Acid Green 5, C.I. 42095). An undyed white cotton sample served as a control.

The dyed cotton mock-ups were subjected to 203 h of artificial photoaging under outdoor light exposure. Regular spectroscopic measurements tracked the rate and nature of colour loss, permitting visual fading with chemical changes to be investigated in the synthetic dye molecules.

By combining historically faithful fabrication with controlled artificial aging, these mock-ups provide a reproducible way to evaluate how natural and synthetic dyes respond to photoageing under both indoor museum lighting and outdoor conditions, covering the full relevant spectral range from UV to visible light. The integration of analytical techniques both non-invasive and laboratory-based ensures a detailed understanding of degradation processes.

This methodology not only safeguards the original textiles but also produces data directly applicable to conservation decision-making, including the optimisation of lighting protocols and storage conditions. Ultimately, the study of these mock-ups contributes to the development of evidence-based preventive conservation strategies for both natural-dye and synthetic-dye textiles.

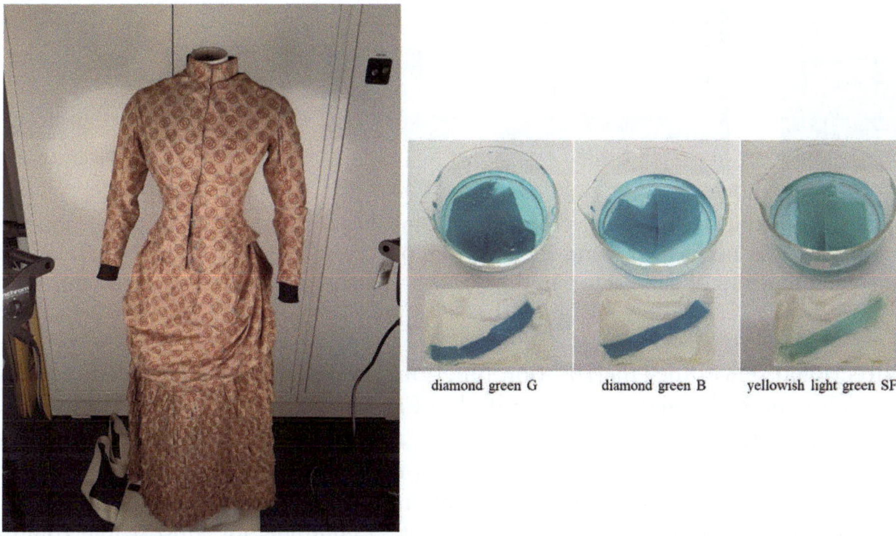

Fig. 27. Left: Victorian printed cotton dress (Edmund Potter & Co., 1887), Victoria and Albert Museum (T.2-1926), showing fading in the background. Right: Dyed cotton mock-ups prepared to match the original dress's colours, for use in artificial aging experiments.

4.3 SVBRDF Capture and BRDF Modelling

To accurately capture the SVBRDF of our mockup samples, we utilize the Total Appearance Capture (TAC7) [14,20,35], developed by X-Rite. This commercially available scanner is specifically designed for capturing material appearances and operates with Pantora software [36], which facilitates postprocessing and optional editing of the resulting SVBRDFs.

The TAC7 features a hemispherical array comprising 29 white LEDs and four monochrome cameras arranged in an arc above a rotating turntable. Five of the LEDs are equipped with colour filter wheels that include 10 filters covering the visible spectrum. To ensure detailed sampling of reflectance lobes, a linear light source can be rotated in the space above the sample holder. For each material analyzed, we gather 100 colour images, 348 monochrome point-lit images, and 280 linear light source images [19]. Additionally, a structured light projector is employed to capture low-resolution surface geometry, and transparent materials are illuminated from below using an LED that passes through a diffuser plate in the sample holder. The TAC7 generally requires about 30 min to process low-gloss anisotropic fabrics [19]. While it can detect fluorescence and alerts users to its presence, it does not incorporate fluorescence into the measurements; however, a mask can be used to identify sections of fluorescent fabric.

The captured appearance of textile mockups (kimono silk and Victorian dress) using the TAC7 device was used to extract the material maps with Pantora [36]. This software fits the appearance data to a spatially varying

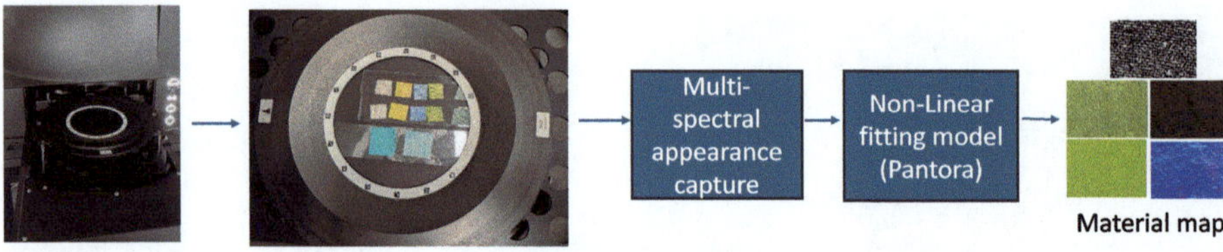

Fig. 28. Appearance capture was conducted using TAC7 [14], and fitting to the appearance model was performed using Pantora [36] on the silk mockup of the kimono and the cotton mockup of the Victorian dress, both created by a conservation scientist.

bidirectional reflectance distribution function (SVBRDF) and generates the corresponding material maps, as illustrated in Fig. 28. To the best of our knowledge, the fitting model provided by Pantora is regarded as state-of-the-art for achieving high-quality material fits. The anisotropic SVBRDF obtained from a TAC7 measurement consists of 13 channels: 3 channels for diffuse colour, 3 channels for normal maps, 3 channels for specular colour, 3 channels for anisotropic roughness, and 1 channel for Fresnel effects (see Fig. 28).

4.4 Fading Simulation

Our approach to simulating colour fading relies on two main assumptions: either we have access to mockups that accurately replicate both the visual appearance and chemical composition of the dyes present in the studied textiles, and that are subjected to controlled accelerated aging, or we possess a series of visual records (photographics documentation) that document the objects' colours over time. The spectral reflectance measurements collected from the mockups at each stage of the accelerated aging process can be integrated with the appearance modeling framework outlined in the previous section, enabling a continuous simulation of colour changes in the textiles.

By measuring how light reflects off the surface of these mockups at different stages of aging, we can build a model that shows how the colours in the original textiles progresses throughout the aging timeline. This allows us to "rewind" the visual appearance of an object to its original state and then "play forward" its gradual fading, almost like watching a time-lapse of its life. This method presumes that the mockups fully and faithfully represent the appearance of the actual objects, which is rarely the case. As we can see in Figs. 29 and 30, the colours resulting from the direct appearance transfer from the mockups, although plausible, are rather saturated and vivid, in stark contrast with the real heritage object that is characterized by a natural patina and softer colours.

To improve accuracy and convey a more natural looking effect, we compare the mock-ups' appearance with the current condition of the textiles using hyperspectral imaging. Thus, for each textile, we capture two details with a hyperspectral camera. The details were selected in such a way so that they contain the fugitive dyes in a good condition. With the spectral unmixing technique,

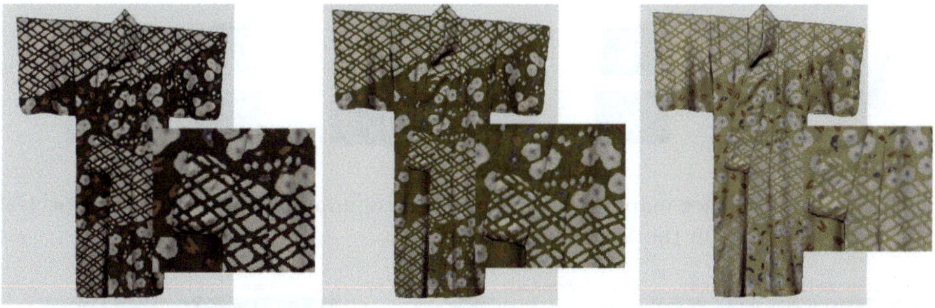

Fig. 29. Past and future simulations of the kimono in Fig. 21, generated with direct appearance transfer from the mock-ups: unaged with reduced exposure (left), unaged as recorded by TAC7 (b), and aged as recorded by TAC7 (c). All renderings utilize a spotlight source with a colour temperature of 6600K.

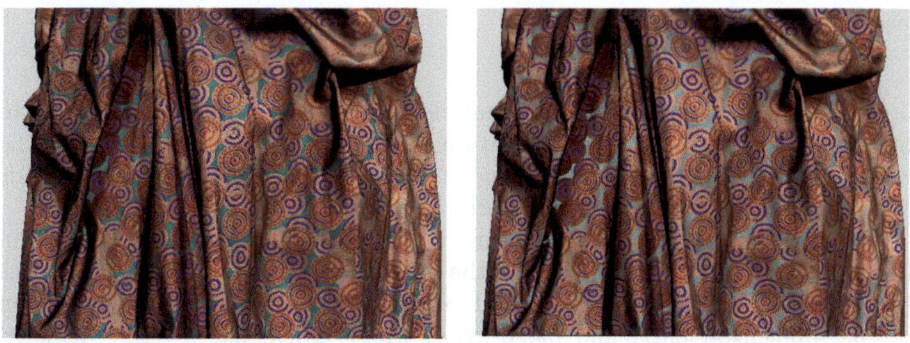

Fig. 30. Past and future simulations of the turquoise dye in the Victorian dress in Fig. 21, generated with direct appearance transfer from representative mock-ups: unaged as recorded by TAC7 (a), and aged as recorded by TAC7 (b). Additionally, the faded violet regions have been rendered with a representative colour derived from analytical studies. All renderings utilize a spotlight source with a colour temperature of 6600K. (Color figure online)

we then break down the hyperspectral data into the constituent pure materials in the object, known as endmembers. For materials recreated as mock-ups, the spectral signatures of the endmembers are derived directly from the mock-ups; for those not represented by mock-ups, the signatures are extracted from the hyperspectral images themselves. Following spectral unmixing, we generate abundance maps that indicate the spatial distribution of each dye across the textile surface (Fig. 31).

These maps can then be recombined with either aged or unaged spectral data from the mock-ups to simulate the fading process within the hyperspectral image, offering a dynamic and data-driven visualization of colour degradation over time. For the Victorian dress, we first define the turquoise endmember with the spectral signature of the corresponding unaged mock-up (top row in

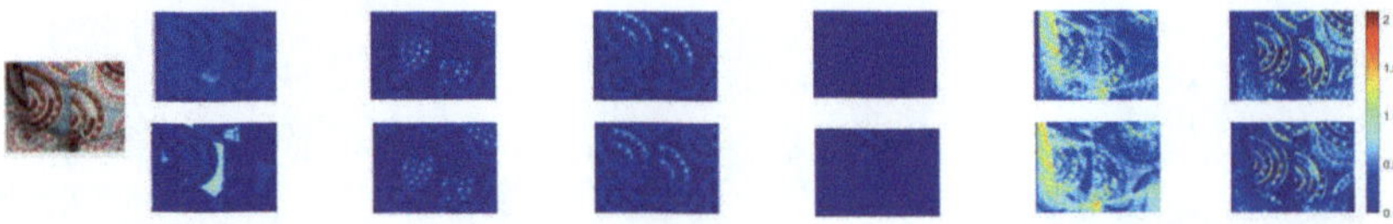

Fig. 31. Dye abundance maps obtained with spectral unmixing from the hyperspectral images of the details in the Victorian dress.

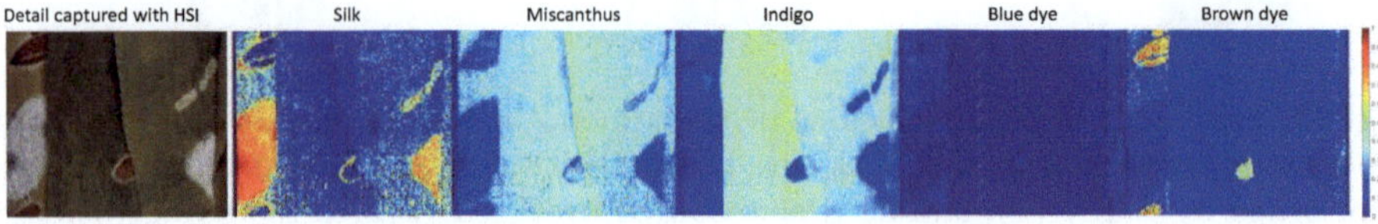

Fig. 32. Dye abundance maps obtained with spectral unmixing from the hyperspectral images of the details in the kimono.

Fig. 31). This is used for fading simulation from present to future. Then, in the bottom row, we extract the signature of the turquoise endmember from the hyperspectral image itself, as this dye is applied purely and unmixed. This is used for attempting to virtually restore the original colour in the dress.

The green in the kimono is a mixture of miscanthus and indigo, so we perform the unmixing with the two spectra of the dyes taken from the unaged mock-ups. Because the detail scanned with HSI contains both green in good condition (in the top-central part) and faded green where indigo has severely depleted (left margin), we attempt the restoration of the past colors with former signal and the present to future discoloration with the latter signal. This simulation correlates any observed colour changes with actual chemical alterations, enabling a meaningful assessment of the progressive loss of colour that is directly linked to the behavior of the original materials. As a matter of fact, the simulations look more natural compared to the mock-up appearance mode, as can be noticed in Figs. 33 and 34.

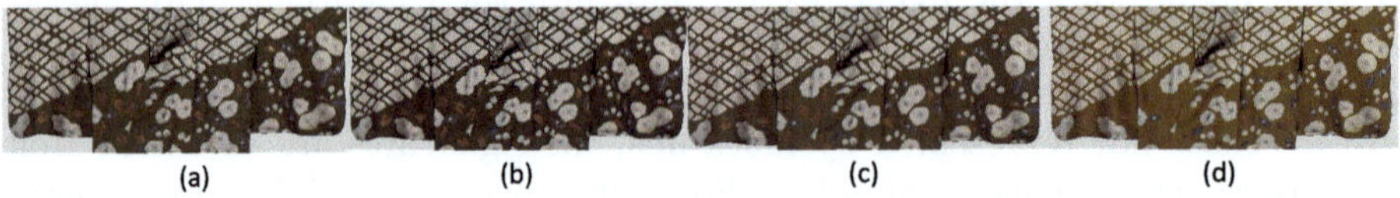

Fig. 33. Fading simulation for the kimono at various time intervals – hypothesized original colour (a) and fading expected in the future (b, c and d) – considering correlation of the mock-ups with the actual heritage clothings through spectral unmixing.

The reconstruction of the kimono using photographic documentation (Fig. 35) reveals significant changes in color and detail from the front view in

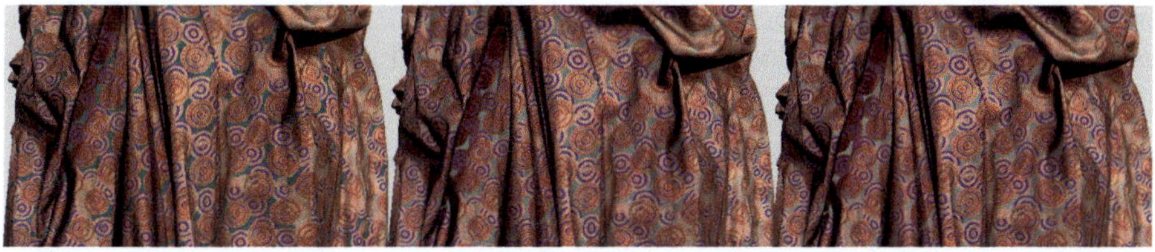

Fig. 34. Fading simulation for the turquoise dye in the Victorian dress at various time intervals – hypothesized original colour (left) and fading expected in the future (middle and right) – considering correlation of the mock-ups with the actual heritage clothings through spectral unmixing.

2006 (a) to the front view in 2019 (b), highlighting the effects of time on its appearance. Likewise, the inner surface of the kimono (c), photographed in 2024, exhibits less fading compared to its current appearance captured in the 3D model (see Fig. 21), providing valuable insights into how the colors may have originally appeared. However, the reconstruction based on the reverse side of the kimono appears to be intermediate between the 2006 and 2019 reconstructions. This is expected, as the inner surface, although shielded from light exposure, likely experienced discoloration due to other untracked aging factors such as humidity and temperature affecting the inner folds. Similarly, when comparing the inner surface reconstruction of the Victorian dress (Fig. 35(d)) with its present-day appearance captured in the 3D model (Fig. 21), it is evident that the former is less faded than the latter.

Reconstructing color changes from chronologically archived photographs allows for retrospective modeling of degradation, offering important qualitative insights into the temporal evolution of dye fading and the long-term behavior of historic textile materials, though it does not provide evidence of the environmental factors influencing color alterations. A significant limitation of using documentation pictures taken over various years is the ambiguity introduced by

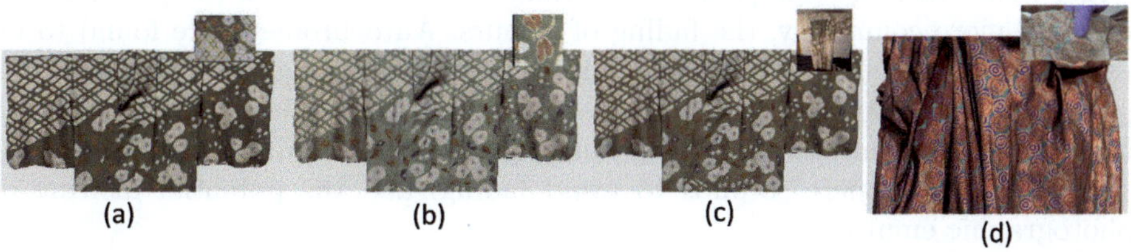

Fig. 35. Reconstruction of the kimono's historical colors using reference documentation images shown in the top corners: (a) from 2006, (b) from 2019 and (c) depicting the inner surface of the kimono that was shielded from light exposure. Also included is the Victorian dress (d) based on a photograph from the inner surface. All images are illuminated with a spotlight set to a colour temperature of 6600K.

subjective choices in photography, such as differing view and illumination setups, as well as potential variations in post-processing and color calibration pipelines.

5 Modelling Colour Change in Historical Photo/Film Collections

The investigation of the deterioration of autochromes and Kodacolor films and the reconstruction of their original appearance are essential to appreciate the full scope and impact of these pioneering colour processes. While these early technologies represent significant milestones in the history of photography and cinematography, their unique aesthetic qualities are often lost due to their inherent fragility, chemical instability and physical degradation.

By digitally reconstructing their original colours and appearance, we make it possible for the general public to experience these works as they were intended—vivid, immersive, and historically rich—thereby fostering greater appreciation and enjoyment of early colour imaging, its cultural significance and its societal impact.

An Introduction to Autochromes

The Autochrome Lumière is considered the first commercially successful colour photography process. Publicly released in 1907, the invention by the Lumière Brothers was publicly praised for its production of luminous colour photographs. Autochromes comprise a glass plate, on top of which sits thousands of starch granules dyed red, green and blue, which act as a colour filter. On top of the colour filter is a layer of panchromatic silver emulsion. In essence, the colour filter produces the impression of colour in the photograph, while the silver gelatin emulsion provides the image, or outline.

Autochromes were used primarily in the period 1907 to 1930. During this time, multiple problems with the process became apparent, particularly in regards to the presentation and preservation of colour. Articles in photographic journals from the period discuss the presence of issues including 'greening', which manifest on the autochrome plate as small spots or large areas in shades of green. Other issues included the presence of orange areas or spots on the plate, silvering and, more commonly, the fading of colours. Autochromes were found to be extremely light sensitive, and popular means of exhibiting glass plates in the period like projection caused the colour dyes to quickly fade. Articles dating from the early twentieth century discuss limiting projection and the light exposure of plates to mere seconds to avoid fading, and the potential melting of photographic emulsion.

Today, as well as in the early twentieth century, it was understood that when the colour in autochromes changed or was damaged, there was little action that could be taken to restore the plates to their original appearance. PERCEIVE has worked on new initiatives that offer means to generate an image which provides a sense of what the autochrome may have originally appeared like.

An Introduction to Kodacolor

Since the emergence of cinema in the 1890s, inventors and researchers sought ways to bring colour photography into motion pictures. However, it took decades before a workable method of recording colour in film was achieved. The lenticular process offered a solution by capturing three colour separations simultaneously within a single film frame. The principles behind this technique originated from earlier work by pioneers such as Liesegang and Lippmann. Liesegang had imagined a pixelated approach using a perforated screen, while Lippmann described in detail the use of lenticular film to achieve more realistic colour reproduction [22]. In the early 1900s, Berthon built upon these ideas and applied them to moving pictures [3]. It took a long time before Kodak managed to refine and commercialise the technology quickly, launching Kodacolor in 1928 as a 16 mm reversal film designed for amateur use. For any new colour process to thrive, it needed widespread adoption, compatibility with existing recording and projection equipment, and the ability to produce distributable copies. Kodacolor struggled in the latter area. Moreover, it suffered from the common drawbacks of additive colour methods, including low image brightness, reduced sharpness, and a limited colour range. Although Kodacolor enjoyed some popularity among amateur filmmakers, its appeal quickly faded when Kodak introduced Kodachrome, and lenticular film disappeared by the late 1930s.

Today, many lenticular films remain hidden or misidentified, as they appear to be ordinary black-and-white reels at first glance. Raising awareness of these films is crucial to ensure their proper identification, preservation, and eventual rediscovery for future generations.

5.1 Modelling Colour Change in Autochromes

We present two distinct methodologies for the colour restoration of autochromes, both of which follow a shared processing pipeline illustrated in Fig. 36, previously published in [5]. This pipeline begins with the separation of the colour mosaic from the silver emulsion and proceeds by replacing the silver-decoupled dyes with unfaded reference dyes to achieve colour restoration. The key difference between the two approaches lies in the strategy employed for silver/dye separation. In this section, we will only focus on one method: spectral unmixing. For more details about the imaging set-up and silver/dye separation, please refer to the original publication.

This method capitalizes on the spectral resolution capabilities of our imaging system by formulating the silver/dye separation as a spectral unmixing problem. Following the unmixing process, we replace the spectral signatures of the degraded dyes in the dye mosaic with those of unfaded reference dyes. The restored colour mosaic is then recombined with the silver emulsion. To the best of our knowledge, this represents the first application of spectral unmixing techniques to the restoration of autochromes.

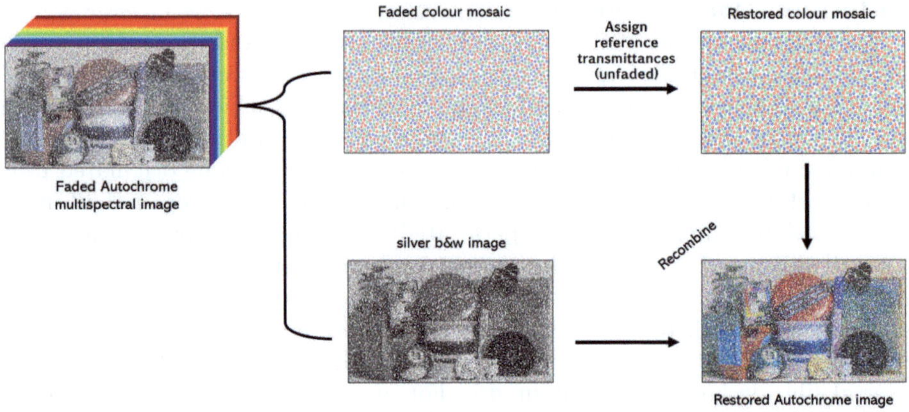

Fig. 36. Our restoration pipeline is defined by the processing of the colour mosaic after its uncoupling from the silver emulsion layer.

Test Autochromes. The significant light sensitivity of autochromes means that many museums choose to limit public access and restrict exhibiting original plates. To avoid degradation on autochromes from public collections, we purchased four autochromes from antiques sites and private collectors for the purpose of the project, each exhibiting varying degrees of fading and typical deterioration such as greening (Fig. 37).

High Resolution Multispectral Image Capture. To capture the fine structural and spectral details of autochromes, we employed a high-magnification multispectral imaging system. Autochromes are composed of microscopic dyed potato starch granules, each approximately 10 μm in diameter. To resolve individual granules, our system achieves a spatial resolution of 1.84 μm on the object plane, enabling precise analysis of the colour mosaic.

The imaging setup is a modified version of the active multispectral system described in [32], consisting of a monochrome 16-bit CCD camera paired with a 5$\times$ magnification lens (Fig. 38a). Illumination is provided by an integrating sphere equipped with ten LEDs, each emitting at distinct spectral peaks in the visible spectrum.

To ensure uniform illumination, the light from the integrating sphere is collimated using a condenser lens and diffused through a translucent slab, creating a bright-field setup. The autochrome plate is placed on a custom holder with a black cover (Fig. 38b), exposing only the area being imaged to minimize stray light and reduce unnecessary exposure. The holder is mounted on an XYZ translation stage, allowing for precise positioning and tile-based image acquisition, which can later be stitched into a complete image.

Exposure times are optimized individually for each spectral band to maximize the dynamic range of the 16-bit sensor without saturation. Due to the inherently low transmittance of autochrome glass plates [10], we use a neutral density (ND)

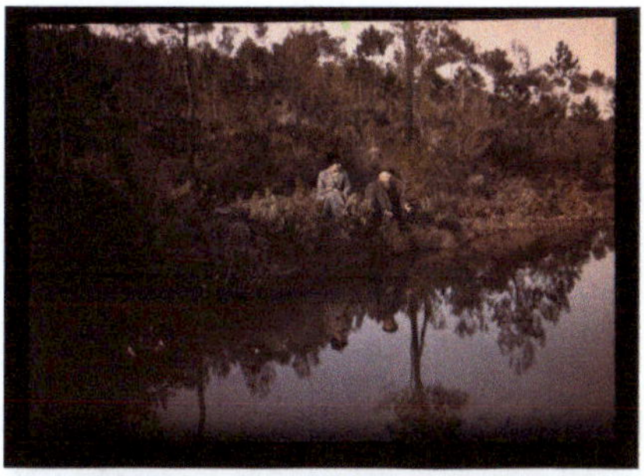

(a) Couple by the lake. Autochrome used by permission of the collector, Arthur Clay.

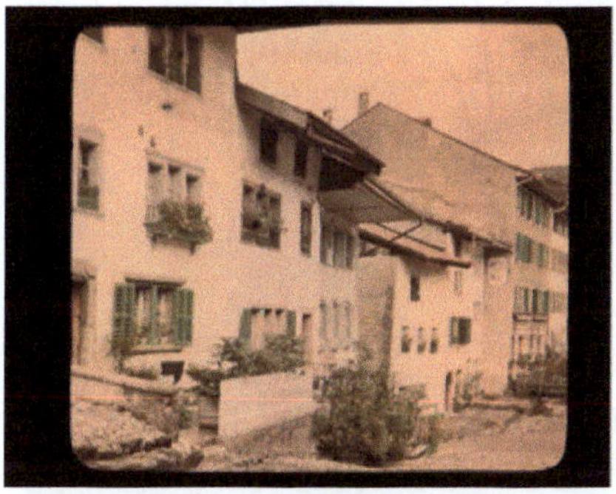

(b) Gruyere, 1914

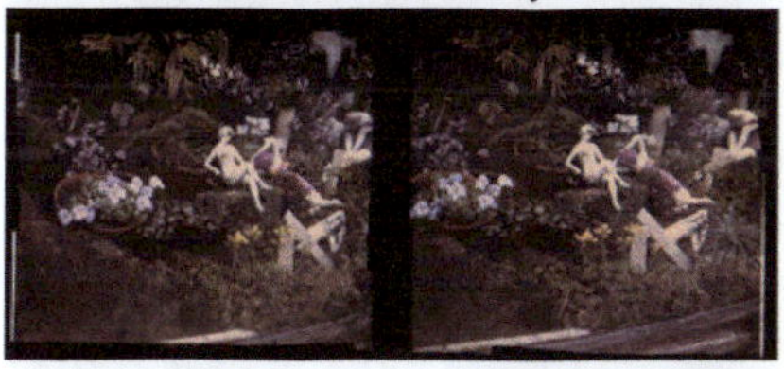

(c) Harrogate shop window

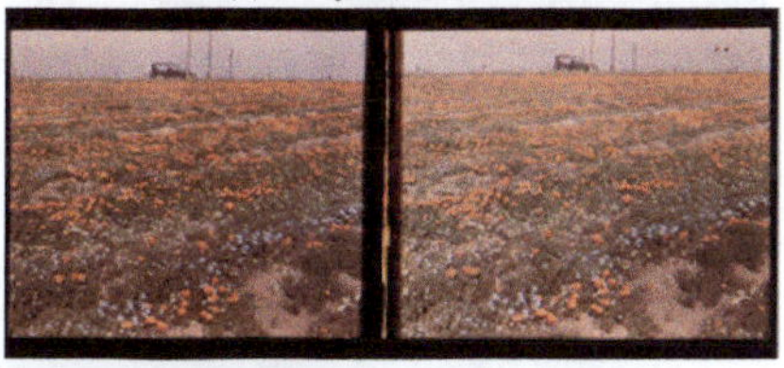

(d) Poppy field

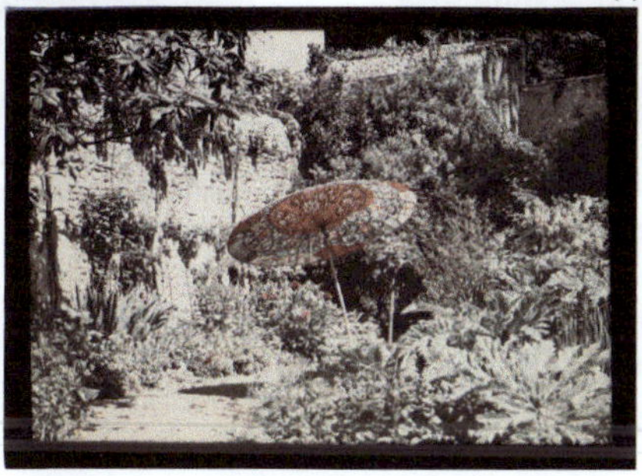

(e) Umbrella. Autochrome used by permission of the collector, Arthur Clay.

Fig. 37. Test autochromes acquired for this research. (Images not to scale)

filter with 0.4 optical density as the white reference, rather than an empty gate. Each captured image undergoes flat-field and dark-current correction.

Given the extreme light sensitivity of autochromes, we also assessed the potential impact of our imaging process on material degradation. This monitoring enables us to compare current and future captures, providing a framework to evaluate any fading induced by the digitization process.

Spectral Unmixing. In our spectral unmixing approach to autochrome restoration, we treat the three dyes—red-orange (R), green (G), and blue-violet (B)—as endmembers. Silver and carbon, which primarily modulate image contrast rather than contribute distinct spectral signatures, are captured in the residuals of the unmixing model. This rationale allows us to focus on separating

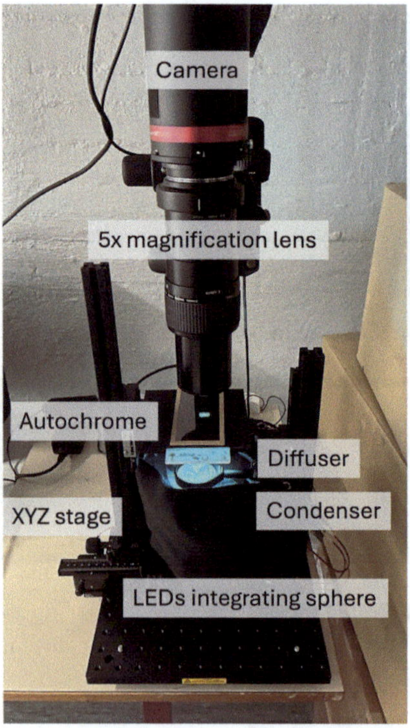

(a) Multispectral Imaging Setup.

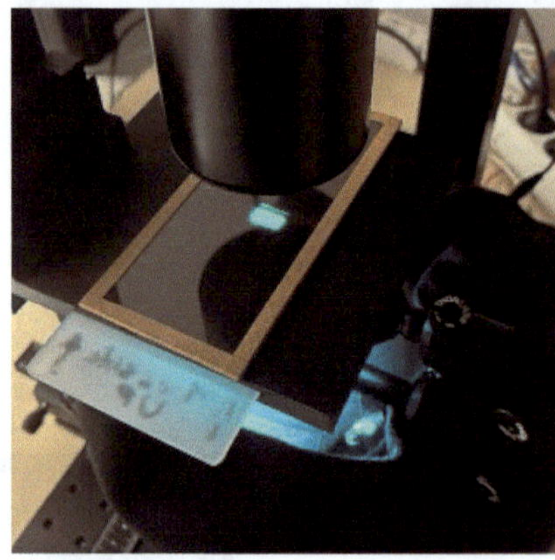

(b) Setup closeup - Autochrome plate on holder, with black cover to minimize stray light and exposure.

Fig. 38. Our autochrome digitization setup is non-invasive and offers flexibility in adjusting the distance between the photograph and the camera.

the dye components from the monochromatic signals introduced by silver and carbon.

For each autochrome, we construct a multispectral image cube using only the bands within the visible spectrum. To identify the spectral signatures of the dyes, we select a region of the image that is free from silver interference. Within this no-silver area, we sample the spectra of individual dye granules and average them to obtain representative spectral profiles for each dye's optical density.

Operating in the additive absorption domain, we assume a linear spectral mixing model. That is, each pixel ij in the multispectral cube is considered a linear combination of the three dye endmembers:

$$OD_{cube} = ab_R^{ij} * OD_R + ab_G^{ij} * OD_G + ab_B^{ij} * OD_B + residual^{ij} \quad (5)$$

with $i = 1...N$ and $j = 1...M$, where N is the number of rows and M is the number of columns in the multispectral image. This assumption simplifies the unmixing process and enables us to estimate the spatial abundance of each dye across the image, forming the basis for accurate colour restoration.

The objective of the spectral unmixing process is to estimate the abundance ab of each dye at every pixel (i, j) in the multispectral image cube. These abundance values represent the spatial distribution of each dye across the image, forming what is known as an abundance map. While abundance is proportional to dye concentration, it does not correspond to absolute quantities and lacks a standardized unit of measurement.

To compute the abundance maps, we solve the linear mixing model using a least-squares minimization approach [13]. The optimization minimizes the squared difference between the observed multispectral cube and its reconstruction based on the linear combination of the dye endmembers:

$$|OD_{cube} - \widehat{OD}_{cube}|^2$$

This minimization is subject to the constraint that all abundance values remain non-negative, ensuring physical plausibility. The resulting abundance maps provide a robust estimate of the spatial presence of each dye, enabling accurate colour restoration while implicitly accounting for the influence of silver and carbon through the residuals of the model.

Once the abundance maps for each dye are computed, we reconstruct the multispectral image cube by multiplying these maps with the corresponding spectral signatures of the dyes:

$$\widehat{OD}_{cube} = ab_R^{ij} * OD_R + ab_G^{ij} * OD_G + ab_B^{ij} * OD_B. \tag{6}$$

This yields a modelled version of the original multispectral cube, representing the contribution of the dye mosaic alone.

The residual error is then calculated as the difference between the original multispectral cube and the reconstructed cube:

$$residual^{ij} = OD_{cube} - \widehat{OD}_{cube} \tag{7}$$

This residual captures the spectral information not explained by the dye components—primarily the influence of the silver emulsion and carbon elements. In this way, the abundance maps effectively isolate the spatial distribution of the dyes, while the residual serves as a segmentation of the silver layer and other non-dye components.

Colour Restoration. To restore the original colours of the autochrome dyes, we rely on the spectral transmittance measurements of unfaded dyes previously obtained by Lavédrine and Gandolfo [16] from a well-preserved historical autochrome. These spectral profiles represent the optical characteristics of the red-orange, green, and blue-violet dyes used to colour the potato starch granules.

Using the spectral unmixing approach, we decompose the multispectral cube of each autochrome into abundance maps for the three dyes, as described by Eq. 5. Faded dyes typically exhibit lower and flatter optical density curves compared to their unfaded counterparts. To reconstruct the unfaded appearance, we multiply the abundance maps by the reference optical densities of the unfaded dyes, resulting in a restored colour mosaic.

To reintroduce spatial detail, we add the residual component which captures the silver and carbon contributions back to the restored mosaic. This yields the final restored multispectral cube, expressed as:

$$OD_{restored\ cube} = ab_R^{ij} * OD_{Ref\ R} + ab_G^{ij} * OD_{Ref\ G} + ab_B^{ij} * OD_{Ref\ B} + residual^{ij} \quad (8)$$

The restored cube is then rendered using the CIE 1931 2° standard observer and the D50 standard illuminant. For final image encoding, we apply the Rec. 2020 colour profile, which follows the ITU-R BT.1886 [25] recommendation and includes gamma encoding with $\gamma = 1/2.4$.

Colour Restoration of Partially Faded Autochromes. The test autochromes were imaged using the high-resolution multispectral setup previously described. Figure 39 presents a selected area from each autochrome where the silver emulsion is assumed to be transparent. The spatial resolution of the imaging system is sufficient to resolve individual dyed starch granules, allowing for a detailed assessment of their current state of conservation. The optical densities of the dyes were computed by averaging pixel values within the selected area shown in Fig. 39. Compared to the reference optical densities of autochromes, the faded dyes exhibit lower maximum densities, indicating reduced light absorption, and higher minimum densities indicating reduced light transmission. This dual effect results in diminished contrast and desaturated colours, characteristic of aged autochromes.

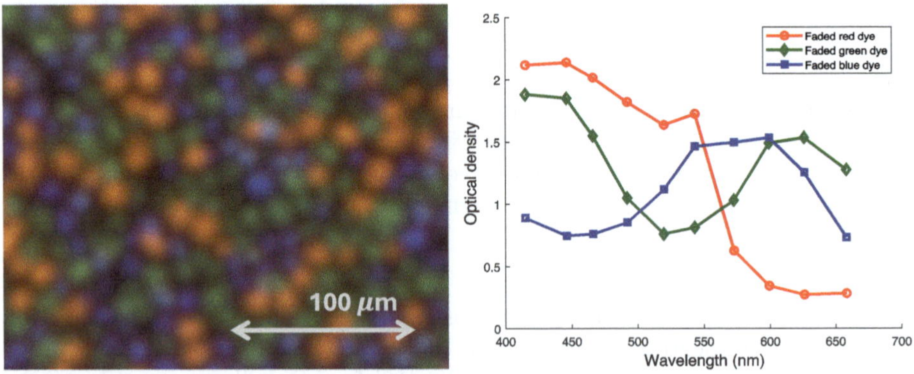

Fig. 39. Left: Dyes uncovered by the silver emulsion, cropped from the colour image of the autochrome Couple by the lake. Right: Spectral signatures of the pure dyes in the "Couple by the lake" autochrome.

For each autochrome in our study, we selected a region devoid of silver emulsion which served as the basis for extracting the average spectral signatures of the red, green, and blue dyes within the visible range Fig. 39. These spectral profiles were used as endmembers in the spectral unmixing method, allowing us to compute the spatial distribution of each dye across the entire image in the form of abundance maps.

Beyond dye separation, we also decomposed the silver and carbon signals. Because carbon is optically mixed with the adjacent dyes, its contribution is embedded within the abundance maps. The silver layer, however, is isolated by calculating the residual between the original multispectral image and the image reconstructed using the unmixing model.

To reconstruct the original multispectral cube, we multiply the abundance maps by their corresponding endmember spectra, accounting only for the dyes and carbon elements. The resulting colour rendering of the faded dye mosaic is shown in Fig. 40b. A faint ghost-like artefact of the woman's silhouette is only slightly visible, likely due to scattering effects not captured by our current model.

For the virtual restoration, we replace the faded dye spectra with the reference spectra of unfaded dyes, combining them with the abundance maps to generate a restored dye mosaic (Fig. 40d). Finally, we add the silver layer, represented by the residual image (Fig. 40e) to the restored dye mosaic, producing the fully restored autochrome shown in Fig. 40c.

Colour Restoration of Autochromes with Completely Faded Blue Dye. The autochromes previously discussed exhibit varying degrees of fading across their dyes. In this subsection, we focus on a particularly degraded autochrome in which the blue-violet dye has completely faded, namely the "Umbrella" autochrome shown in Fig. 37e. Figure 41 shows a magnified view of its dyed potato starch granules, revealing that the blue dye granules are nearly transparent, while only the red-orange and green dyes remain visibly present. Further analysis of the spectral curves confirms this observation: not only is the blue dye absent, but the remaining red and green dyes also show significant fading. Their optical density curves are both flattened and reduced in magnitude, indicating diminished absorption and transmission properties. This results in a severely desaturated colour mosaic with low contrast, characteristic of advanced dye degradation in autochromes.

We applied the spectral unmixing approach, following the same procedure previously outlined. We extracted the spectral signatures of the faded dyes directly from silver-free regions of the image. This allowed us to estimate the spatial distribution of the remaining dye components through abundance maps, even under conditions of extreme fading. Although the spectral contrast between the dyes was minimal, the unmixing method proved sufficiently robust to isolate the residual dye information, as shown in Fig. 42. The resulting abundance maps enabled us to reconstruct the faded dye mosaic and proceed with virtual restoration, despite the near-complete loss of the blue-violet dye and the significant fading of the red-orange and green dyes.

To capture the full extent of the umbrella depicted in the autochrome and to produce a visually compelling result, we created a mosaic composed of 45

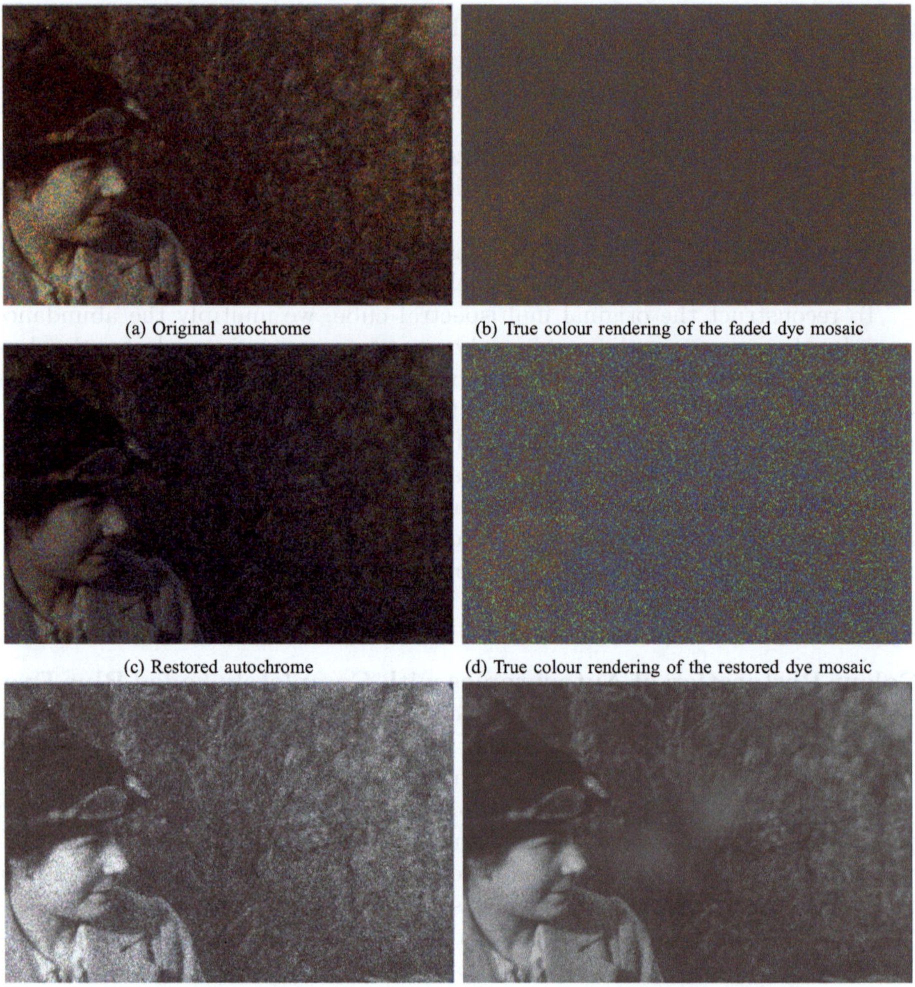

(a) Original autochrome

(b) True colour rendering of the faded dye mosaic

(c) Restored autochrome

(d) True colour rendering of the restored dye mosaic

(e) Residual of the spectral unmixing

(f) Infrared band (940 nm)

Fig. 40. The restoration with the spectral unmixing method (c) pulls up the faded blue colours in the original image, obvious especially in the coat area, and increases the overall contrast. Moreover, the dyes are well decoupled from the silver layer. The restoration is sharper, as the spectral unmixing denoises the image through dimensionality reduction. This is particularly obvious if we compare the silver layer (f) as given by infrared image (that has several artifacts and is affected by blurriness) with the residual of the spectral unmixing (e). (Color figure online)

individual acquisitions. This approach allowed us to preserve fine spatial detail across the entire image while also enhancing its aesthetic impact. The resulting stitched image not only showcases the intricate structure of the dyed starch granules when zoomed in, but also serves as a striking visual demonstration of

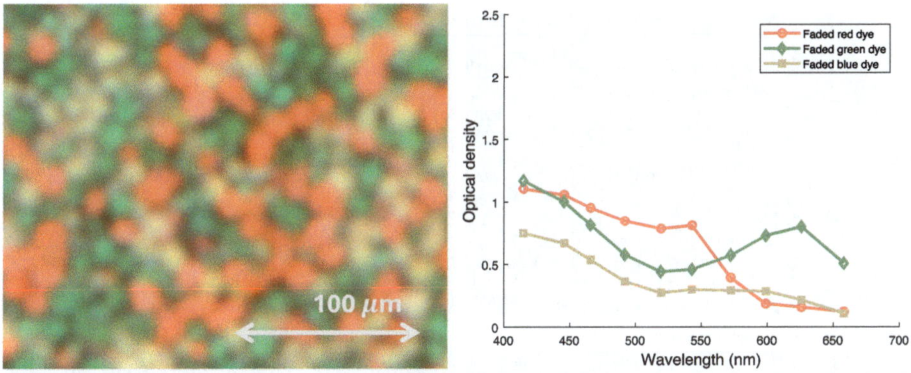

Fig. 41. Left: Dyes uncovered by the silver emulsion, cropped from the colour image of the autochrome "Umbrella". Right: Spectral signatures of the pure dyes in the "Umbrella" autochrome.

the restoration process. By presenting both the original and restored versions, the image effectively illustrates the transformative effect of our method and highlights the contrast between the faded and unfaded colour mosaics (Fig. 43).

5.2 Neural Restoration of Greening Defects in Autochrome Photograph Utilizing Purely Synthetic Data

"Greening" and green spotting (see Fig. 44) are characterized by green areas or spots on the autochrome plate. The historical dyes used to color the starch granules in autochromes are sensitive to water. The green dye, in particular, is vulnerable and can bleed into other areas of the image if moisture penetrates the plate, leading to the greening defect. In severe cases, the entire image can become unreadable.

Since this damage cannot be reversed at all in physical autochrome plates, Sinha et al. [27] proposed using AI models to digitally restore images of digitized autochromes that exhibit greening. To train these AI models for image restoration, we first need to simulate the greening defect, which is referred to as a synthetic defect. The next section describes how we synthetic greening defects [27] were generated.

Synthetic Greening Defect Generation. To address the key issue of insufficient datasets for damaged autochromes, Sinha et al. [27] proposed a method for generating a synthetic dataset suitable for training and testing autochrome restoration techniques. To ensure the quality of this synthetic dataset, we conducted a thorough analysis of actual defects found in autochromes. This analysis revealed that defects can be classified as spot-shaped, widespread, a combination of both, or resulting in complete image loss. We selected seven autochromes for a more detailed evaluation, which demonstrated that spot defects typically

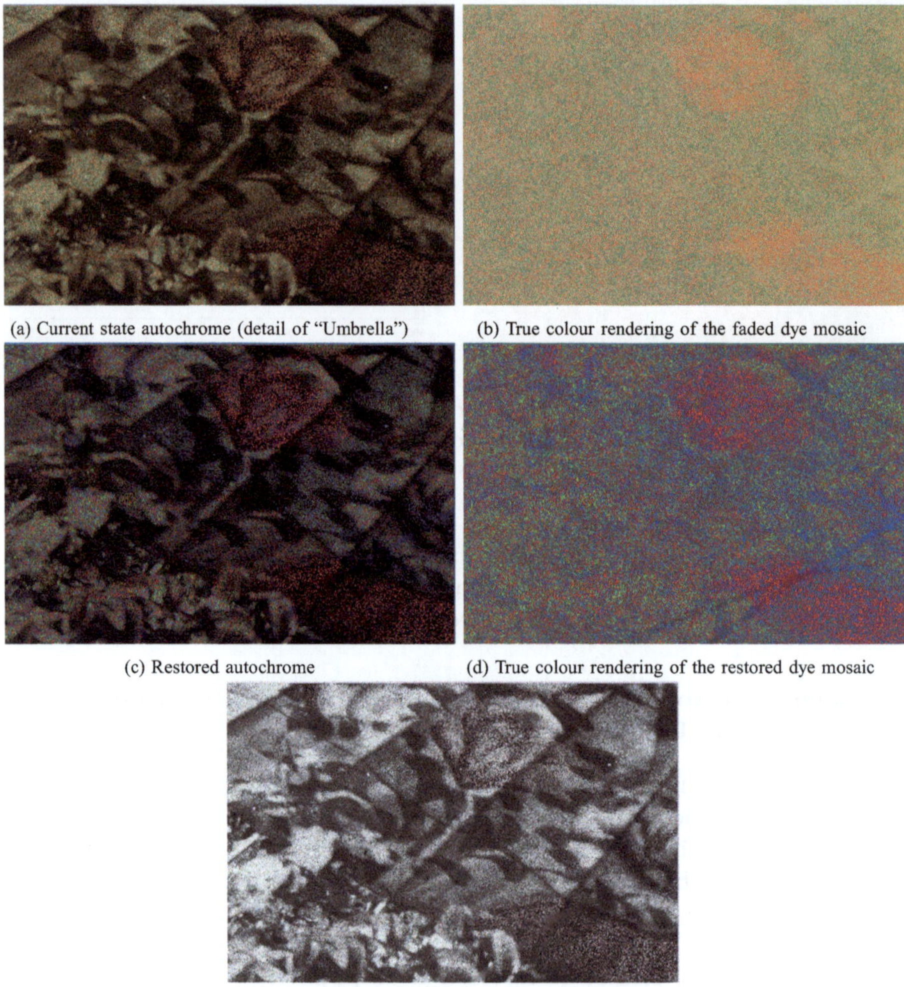

(a) Current state autochrome (detail of "Umbrella")

(b) True colour rendering of the faded dye mosaic

(c) Restored autochrome

(d) True colour rendering of the restored dye mosaic

(e) Spatial information contained by the silver layer, computationally obtained as the residual of the spectral unmixing model

Fig. 42. The "Umbrella" autochrome has no blue colours, because the blue dye has faded completely to the point of becoming white. For this reason, its restoration is more dramatic than in the case of the other autochromes. Apart from the blue dye, the red and green are also rather desaturated, more so than in the other test autochromes. (Color figure online)

range from 1% to 5% of the overall image width, while larger defects can occur independently, covering up to one-third of the image. However, the merging of larger area defects with other spot defects is relatively rare.

In an initially generated dataset, defects were simulated using color filters commonly employed in photography, with a particular focus on green filters to

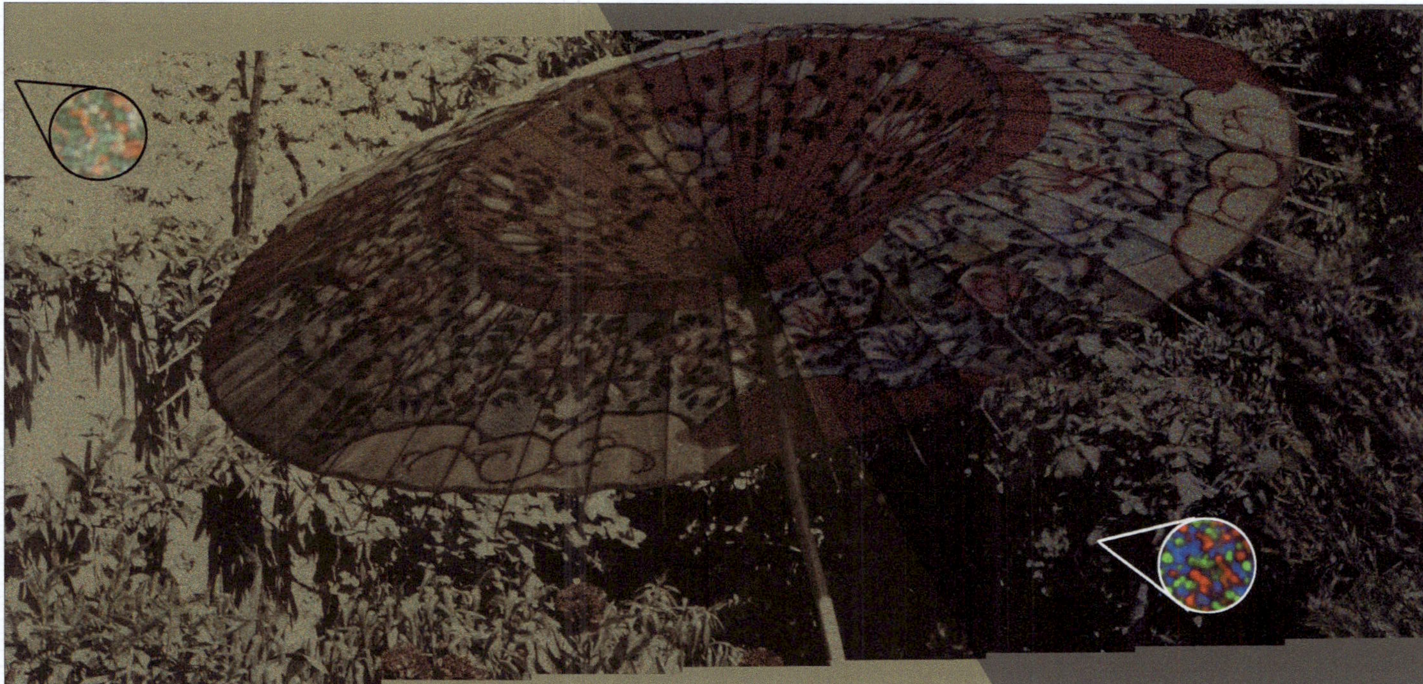

Fig. 43. The full umbrella (as result of stitching 45 captured images) detail in the "Umbrella" autochrome, split between its current, faded state (left side of the diagonal), and its restoration with the spectral unmixing method (right side of the diagonal). We can notice how the umbrella regains not only its blue colour, bleached in the current state, but also the saturation and vividness of its red and green granules. The two magnifier peepholes illustrate the full resolution of the colour granules in the autochrome's mosaic. At this resolution, it is even more obvious how the granules that were once blue are white in the current state as consequence of fading. (Color figure online)

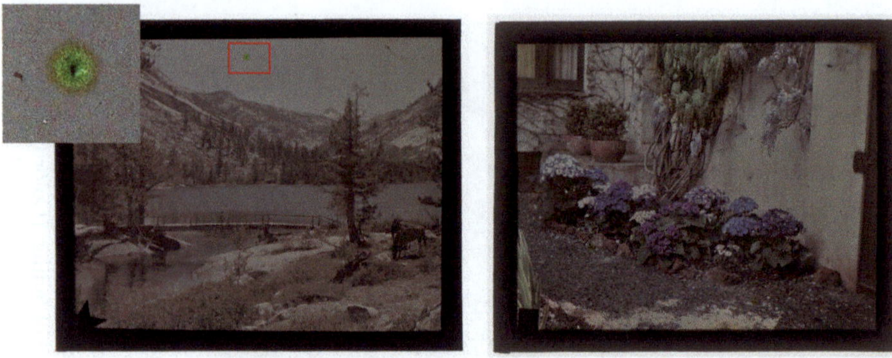

Fig. 44. Examples of real greening defects: small area or spotting defects (left) and large area defects (right). Image source [30].

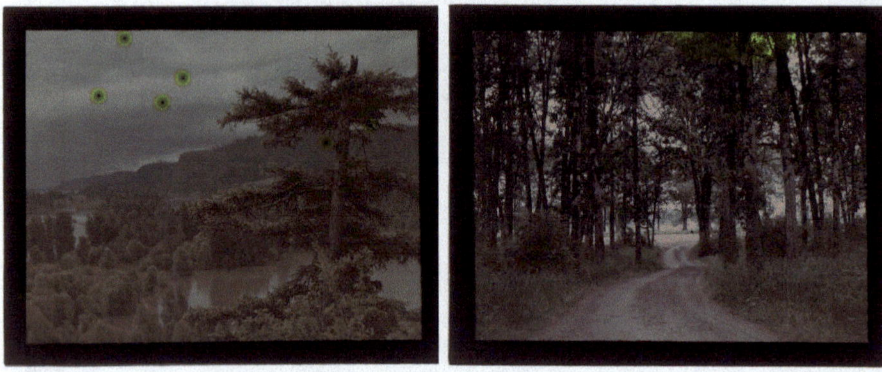

Fig. 45. Synthetic greening defects (small area (left) and large area (right) defect) generated using the approach proposed by Sinha et al. [27]. Image source [30].

represent the foreground layer during the defect generation for the autochrome in the background layer. However, this approach ultimately failed to produce realistic results, as the defects exhibited low color variance, rendering them unrealistic. Hence, a second dataset was created, referred to as ChannelGreeningDefects (see Fig. 45), which directly captures observable changes in the color channels within the defective areas. The process of generating this synthetic defects [27] can be described below:

- Images may contain point defects (60%), large defects (30%), or both (10%), with defect masks generated according to the characteristics of these known defects. These masks are created with 1 to 7 spot-shaped defects or 1 to 2 larger defects, using sizes and centers defined based on known size ratios from real defects.
- The generation of area and spot-shaped defects involves creating ellipses distorted by random radius adjustments at their boundaries. Spot defects have

their centers located within the image, resulting in point-shaped origins, while large area defects are represented by significantly larger ellipses with origins outside the image, simulating the leakage of liquids that penetrate the autochrome and cause damage. The parametric equations [27] for the ellipse are given by:

$$x(t) = x_c + a \cdot \cos(t) \cdot \text{irregularity},$$
$$y(t) = y_c + b \cdot \sin(t) \cdot \text{irregularity},$$

where (x_c, y_c) are the coordinates of the center of the ellipse, (a, b) are the semi-major and semi-minor axis lengths, t is an angular parameter in $[0, 2\pi]$, and irregularity is a factor affecting the shape of the ellipse. The center of defect masks may be point-shaped or linear, and the intensity of the defect is interpolated based on the normalized distance d to the origin, with a decrease in intensity computed by $I = -d^2 + 1$.

- For each ring, a change in the percentage of the individual color channels in the defect areas is defined, randomly adjusted by a factor of 0.2 (chosen empirically to avoid drastic changes and maintain the realism of the defects) in both directions for each image. The combined defect pattern is smoothed using a Gaussian filter, and the intensity change ΔI_c for color channel c is calculated as $\Delta I_c = (p_c * I_c) - I_c$, where p_c is the corruption percentage. This change is then applied to the final image. Table 2 shows the definition of the individual modifications per channel and ring in a dictionary.

Table 2. Dictionary for ring of corruption. Source [27]

Label	Blue	Green	Red	Effect in defect
9	0.6	0.85	1.05	Outer orange ring
1	0.5	1.2	0.8	Light green ring
2	0.4	0.8	0.6	Middle
3	0.4	0.8	0.6	Middle
4	0.2	0.6	0.1	Dark green second
99	0.2	0.2	0.1	Dark mid
20	0.4	0.95	0.6	Surface

Training of Restoration Network. Image2Image models, specifically Pix2Pix [12] and CycleGAN [37], were trained on the synthetic defect dataset (see Fig. 45) using the original training configurations. However, these approaches did not yield satisfactory results in restoring the original colors of the defective areas, as they modified the images without effectively addressing the

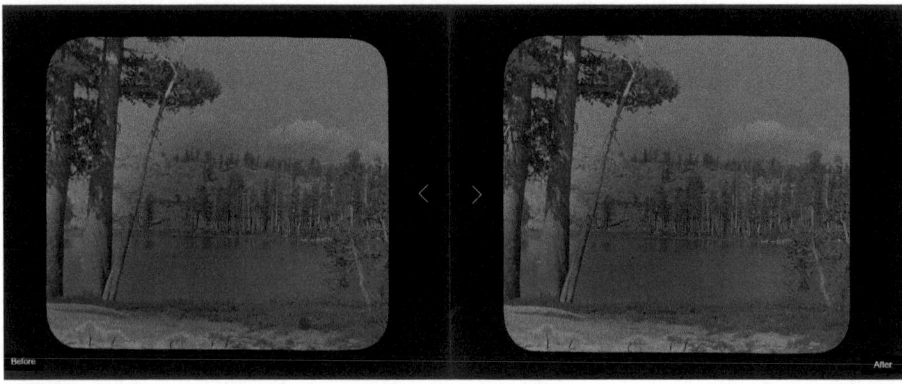

Fig. 46. Large area greening defects removal utilizing the approach proposed by Sinha et al. [27]. Image source [30].

restoration task. Therefore, the ChaIR (Channel Interactions for Image Restoration) model [8] was selected due to its demonstrated state-of-the-art performance in deraining and dehazing tasks which is similar to our problem. This model utilizes spatial and frequency loss functions to evaluate performance and supports dual-channel attention mechanisms. While it primarily affected the greening regions, it showed promise in its application despite the challenges in overall color restoration. The main challenge for autochrome restoration remains the accurate representation of the original colors in the defective areas. To enhance the restoration process, a modified loss function was implemented [27], prioritizing the impact on defective areas over non-defective regions.

A further step involved the digital in-painting of autochromes which have had been de-greened. When digitally restored, the de-greened area can appear pale or off-colour. Using hundreds of undamaged autochromes, we trained the network [15] utilizing low rank adaptation (LoRA) [11] to understand the unique colour palette and texture of the autochrome. Based on this training, the network is able to inpaint those areas digitally "de-greened" so that they more closely resemble the colour and texture of an autochrome (see Fig. 48).

Results. Figure 46 shows that large area defects are removed utilizing this de-greening model. The greening is significantly reduced and this improves the state-of-the-art image restoration methods as evaluated and shown by the work presented by Sinha et al. [27]. Moreover, even greening defects with significant could be removed partially as shown in Fig. 47. In terms of small area defects (green spotting) the de-greened output followed by inpainting using a finetuned model [15] could produce perceptually restored results as shown in Fig. 48. This approach was most effective in stationary regions—areas within an image where statistical properties such as texture, color, and intensity remain relatively constant. This concept is closely linked to stationary texture, which refers to a uniform and consistent pattern throughout a given area.

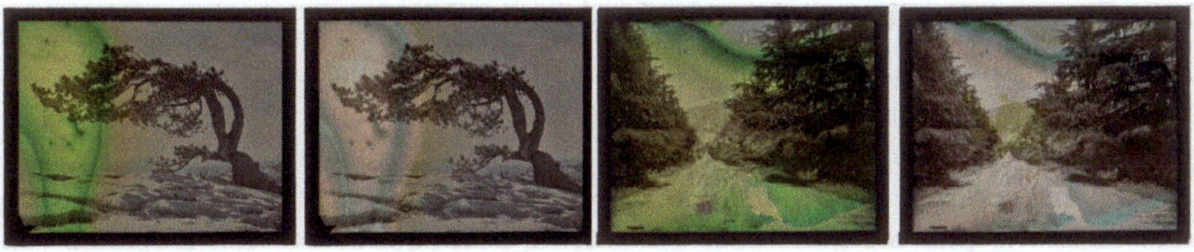

Fig. 47. Large area greening defects (with significant damage) removal utilizing the approach proposed by Sinha et al. [27]. Image source [30].

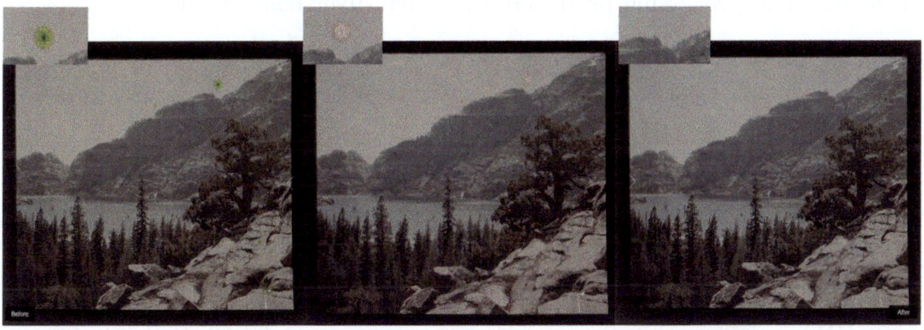

Fig. 48. Removal of small area greening defects (left) using the approach proposed by Sinha et al. [27] (middle), followed by inpainting with a fine-tuned model [15] that employs Low Rank Adaptation [11] (right). Image source [30].

Final Comments. The proposed method effectively identifies and removes some greening defects in old autochrome photographs, regardless of their size or location. By adjusting the colors in the affected areas, the method makes these defects less noticeable. However, it sometimes struggles to accurately reproduce the original colors of the autochrome images, which can lead to bluish tones and the omission of very small defects. To enhance the restoration process, AI-based inpainting methods has been employed to selectively remove small defects using the mask and the de-greened output predicted by the model.

The results of this method can improve editing tools, allowing designers to efficiently achieve similar outcomes while tackling defects that are otherwise difficult to manage, especially in larger areas. Future efforts should aim to expand the synthetic dataset to include digitized autochromes from various collections and to identify effective no-reference metrics—tools that assess image quality without needing a reference image for comparison. These advancements would enable a more comprehensive evaluation of restoration quality across a diverse array of samples. Moreover, the techniques developed could be adapted to address other autochrome defects, such as oranging, and could also be applied to restore various types of artworks and images, broadening their impact in the field of image restoration.

5.3 Reconstructing Colour in Kodacolor Films

Technical Specifications. Lenticular Kodacolor used a form of spatial encoding to capture colour information. The 16 mm film base was embossed with hundreds of fine vertical cylindrical lenses. On the reverse side, a black-and-white panchromatic silver emulsion recorded the image. Figure 49 schematically illustrates the photographic process. During film shooting (left), a special red-green-blue filter (tripartite filter) was placed in front of the camera lens. The lenticular surface focused each colour component onto the emulsion as narrow vertical stripes, enabling colour scenes to be captured on black-and-white film stock. Reversal processing produced a positive silver-based image.

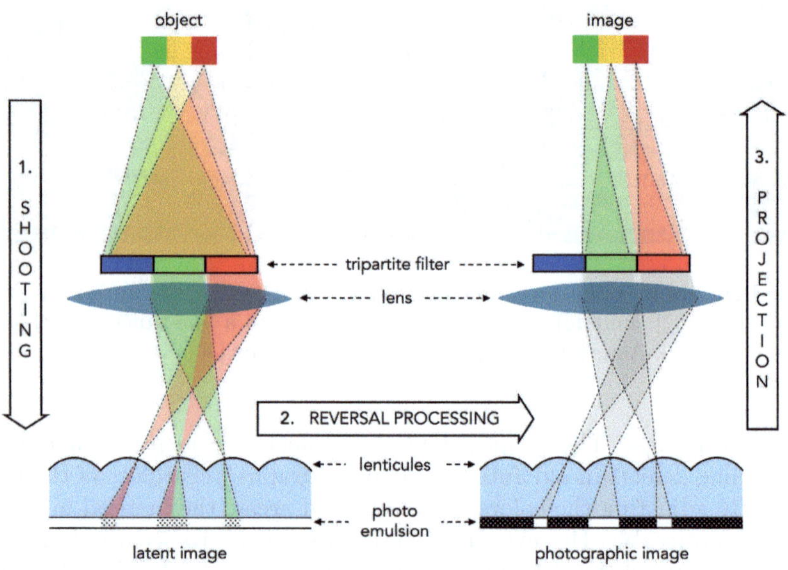

Fig. 49. Schematic illustration of the optical principle underlying image capture and visualization in lenticular colour film, considering a green-yellow-red flag as photographed object. (Color figure online)

The projection (right) required a complementary setup: a projector fitted with a matching red-green-blue filter. When projected, the magnified images recombined to form a full-colour picture on the screen. However, this method of display has since become obsolete. Today, preserving and recovering Kodacolor films requires addressing this issue of technological obsolescence.

Digitisation and Colour Reconstruction. Digitising Kodacolor films provides a practical way to make these works accessible to the public while also safeguarding them against the inevitable decay of the original film material.

Several approaches can be used to extract the embedded colour information. Four main methods can be outlined:

- **Telecine** - This is the most straightforward solution. The original projection system, equipped with the red-green-blue filter, is used, and a digital video camera records the projected colour image on the screen. A synchronization mechanism is required to align the frame rate of the film projector with that of the digital capture device.
- **Colour Digital Capture** - Instead of projecting the image onto a screen (like in "telecine"), a dedicated imaging system can be employed. A lens equipped with a red-green-blue filter–equivalent to the original one–focuses the colour image directly onto a digital colour sensor at low magnification. This process involves frame-by-frame sequential capture using a film transport system.
- **Slit-Scan Capture** - Based on the same principle as *colour digital capture*, the slit-scan eliminates the use of a red-green-blue filter and employs a monochrome digital sensor. A moving slit in front of the lens isolates a specific angular portion of the light, enabling the capture of a single colour component at a time. Three successive exposures per frame, with the slit positioned differently each time, provide the complete colour information. The film transport system is intermittent.
- **Digital Decoding** - In this approach, the film is scanned at high resolution (at 15 or more pixels per lenticule horizontally) with the emulsion facing the imaging system so that the encoded silver-based colour components remain in focus. Specialised software then reconstructs the colour image by:
  1. Locating the lenticular screen, and
  2. Converting the side-by-side silver densities into RGB values.

Among the four methods, *telecine* depends on original Kodacolor projection equipment, which is now extremely rare. Both *colour digital capture* and *slit-scan capture* demand highly specialised, custom-built devices, making them impractical for most institutions. By contrast, *digital decoding* can be considered the most efficient and widely applicable solution. It requires only a conventional high-resolution film scanner for image acquisition, while the reconstruction of the colour information is carried out entirely in software. In 2013, two independent research groups validated the digital decoding approach: Reuteler, Fornaro, and Gschwind within the SNF project doLCE at the University of Basel [24], and Aschenbach in a separate study [2]. The PERCEIVE project draws on the work of Reuteler and Gschwind and on the later advances by D'Aronco and colleagues [9]. The central challenge in digital decoding lies in accurately locating the lenticular screen. The original doLCE method relied on signal processing: it averaged the image rows, then identified the local minima of the pixel value profile to determine the screen's structure. The more recent deep-doLCE approach replaced this strategy with a deep-learning framework. Rather than fine-tuning the existing algorithm with additional parameters, it sought higher robustness and accuracy by letting a neural network learn the structure of the lenticular pattern directly from data.

DoLCE 2.0. Despite the improvement in some cases, deep-doLCE still failed to properly reconstruct some films, therefore, under the auspices of PERCEIVE, the research into this topic has continued. The work focused on a 1930 travelogue to the Norwegian fjords filmed by the amateur filmmaker Arnold Hennings. This 16mm movie is a combination of tinted film (title cards), black-and-white (most of the content), and lenticular Kodacolor (last part). The colour reconstruction using deep-doLCE resulted in image frames that introduced their own set of problems. Many frames displayed artifacts and, in several cases, completely incorrect colours. This issue arises from the lenticule detection model's inability to accurately predict lenticular borders for the provided image set. To improve the colour reconstruction of the travelogue, we refined the original deep-doLCE model using more representative training data aligned with the lenticular boundaries of the scans. This finetuning step enabled the model to more accurately detect boundaries and thereby achieve more reliable colour reproduction. To build this dataset, we employed a semi-automatic labelling process supported by visual analysis, ensuring that reconstructed colours were both technically consistent and historically plausible (e.g., skies appeared bluish, grass greenish, and clothing matched known period tones). We explored two complementary strategies for generating representative data:

- Manual refinement of lenticular boundaries. First, we applied the original deep-doLCE model to colourize the full video sequence. We then selected frames at intervals where colours appeared reasonably plausible. Using these as a starting point, we manually corrected the detected lenticular boundaries to improve accuracy. The corrected frames formed the training data for finetuning the model. If results remained unsatisfactory, additional frames were labelled and incorporated, with the process repeated iteratively until the reconstructed sequence achieved a visually consistent and credible colour balance.
- Automatic selection of plausible frames. To reduce manual intervention, we also adopted a more automated strategy. Instead of adjusting boundaries directly, we identified frames or cropped regions where the original model's output already showed convincing colours. These selections, validated through visual reasoning, were then used to finetune the model. As with the manual approach, additional data were added iteratively until the sequence produced stable and realistic colour results.

Together, these methods demonstrate that user-guided finetuning–whether through manual adjustment or selective frame validation–can substantially improve the colour reconstruction of lenticular film scans, even when source material is of suboptimal quality.

6 Conclusion and Discussion

Colour in heritage objects is both semantically rich and materially fragile. This chapter demonstrates that robust colour reconstruction and change prediction emerge when material science, computational imaging, and visual computing are treated as a single, interoperable workflow. At the foundation are well-designed mock-ups that embody the chemistry and stratigraphy of historical practices—whether Punic-wax polychromy on Pentelic/Parian marble, pigments with linseed oil in paintings, or natural/synthetic dyes on silk and cotton. Colour and spectral imaging, together with appearance capture (MA-T12, TAC7) and SVBRDF/AxF modelling preserve colour, spectral and geometric attributes, supporting photorealistic rendering of instances of the studied artworks in past and future.

classical sculpture, an atlas-based segmentation and semantic shading framework enables the transfer of measured material models from controlled mock-ups onto segmented 3D reconstructions, overcoming earlier "flat colour" restorations by recovering gloss, texture, and micro-geometry cues essential to perceived authenticity of polychromy.

For paintings, we showed that complementary strategies - (i) a mechanistic photochemical model that ties ΔL^*, Δa^*, and Δb^* trajectories to spectral radiant exposure and (ii) spectral unmixing operating in absorbance space - yield predictive maps and renderings of future damage, while historical photographic records can be corrected (via dye density balancing and learned film←painting transforms) to approximate past appearances. This pluralistic approach reduced the typical "lack-of-ground-truth" problem by cross-validating expected behaviours (e.g. vermilion darkening, cadmium yellow fading).

For textiles, integrating spectral unmixing with appearance-aware rendering and interactive visual tools (SimTex) allows conservators to toggle between hypothesized original states and forecasted fading scenarios, grounded in accelerated aging data and archival photographs. The method balances chemical plausibility with perceptual realism, acknowledging uncertainties from photographic variability.

For early colour photographs and films, we advanced multispectral 'silver/dye decoupling' to restore autochromes (including severely faded blue dyes) and explored neural restoration of "greening" defects trained on synthetic data; for Kodacolor, we improved digital decoding through targeted finetuning of deep learning models for lenticule boundary detection, recovering historically plausible colours from lenticular scans.

Author Contributions. Section 2 was written by Yoko Arteaga, Cristiana Barandoni, Donata Magrini, Saptarshi Neil Sinha and Sophia Sotiropoulou. Section 3 was written by Irina-Mihaela Ciortan, Letizia Monico, Irina Sandu, Panagiotis Siozos, and Sophia Sotiropoulou with contributions from David Buti. Section 4 was written by Irina-Mihaela Ciortan, Lucia Burgio, Brenda Doherty, Catarina Monteiro Pinto, and Saptarshi Neil Sinha with contributions from David Buti and Faïçal Selka. Lastly, Sect. 5 was written by Yoko Arteaga, Lucia Burgio, Irina-Mihaela Ciortan, Catlin Langford, Giorgio Trumpy, and Saptarshi Neil Sinha, with contributions from Jon Yngve Hardeberg.

Acknowledgments. This study was funded by the European Union (grant number 101061157, Project PERCEIVE).

Disclosure of Interests. The authors have no competing interests to declare that are relevant to the content of this article.

References

1. Arteaga, Y., Sole, A.S., Hardeberg, J.Y., Boust, C.: Characterising appearance of gold foils and gilding in conservation and restoration. In: Proceedings of the 11th Colour and Visual Computing Symposium (CVCS), vol. 3271. CEUR Workshop Proceedings (2022). https://ceur-ws.org/Vol-3271/Paper11_CVCS2022.pdf
2. Aschenbach, T.: Digital Color Decoding of Kodacolor Film, November 2013
3. Berthon, R.: A suggested new method of colour photography direct at one exposure. Br. J. Photogr. **57**, 421–423 (1910)
4. Chatterjee, S., Trumpy, G., Ruedel, U.: Digital unfading of chromogenic film informed by its spectral densities. Heritage **6**(44), 3418–3428 (2023). https://doi.org/10.3390/heritage6040181
5. Ciortan, I., Arteaga, Y., Trumpy, G.: Silver/dyes decoupling and colour restoration of autochromes using multispectral imaging. Eur. Phys. J. Plus **140**, 384 (2025). https://doi.org/10.1140/epjp/s13360-025-06284-2
6. Ciortan, I.M., Trumpy, G.: FILM2PAINT: transforming photographic documentation on reversal film into paintings' accurate colors. In: Archiving Conference, vol. 21, pp. 17–22. Society for Imaging Science and Technology IS&T, 7003 Kilworth Lane (2024)
7. Ciortan, I.M., Trumpy, G., Sandu, I.C.A.: *The Scream* (ca. 1910) through the Years: from Photographic Documentation to Digital Rejuvenation. In: Colour Photography and Film: Sharing Knowledge of Analysis, Preservation, Conservation of Analogue and Digital Materials, p. 152 (2024)
8. Cui, Y., Knoll, A.: Exploring the potential of channel interactions for image restoration. Knowl. Based Syst. **282** (2023)
9. D'Aronco, S., Trumpy, G., Pfluger, D., Wegner, J.D.: A deep learning approach for digital color reconstruction of lenticular films (2022). https://arxiv.org/abs/2202.05270, eprint: 2202.05270
10. Hannouch, H.: Three-colour photography: technologies, expeditions, empires. Eur. Soc. Hist. Photogr. **37**, 5 (2022)
11. Hu, E.J., et al.: LoRA: low-rank adaptation of large language models (2021). https://arxiv.org/abs/2106.09685
12. Isola, P., Zhu, J., Zhou, T., Efros, A.A.: Image-to-image translation with conditional adversarial networks. In: CVPR, pp. 5967–5976 (2017)
13. Keshava, N., Mustard, J.: Spectral unmixing. IEEE Sig. Process. Mag. **19**(1), 44–57 (2002). https://doi.org/10.1109/79.974727
14. Kohlbrenner, A., Rump, M., Frick, B., Schwartz, C.: Method and apparatus for digitizing the appearance of a real material (2018)
15. Labs, B.F., et al.: FLUX.1 Kontext: flow matching for in-context image generation and editing in latent space (2025). https://arxiv.org/abs/2506.15742
16. Lavédrine, B., Gandolfo, J.P.: The Lumière Autochrome: History, Technology, and Preservation. Getty Publications, Los Angeles, California (2013)

17. Liverani, P., Bracci, S., Iannaccone, R., Lenzi, S., Magrini, D.: New evidence about the polychromy of early imperial cycle from the *Augusteum of Rusellae* (Tuscany). Heritage **6**, 3385–3401 (2023). https://doi.org/10.3390/heritage6040179
18. Magrini, D., Bracci, S., Bartolozzi, G., Iannaccone, R., Lenzi, S., Liverani, P.: Revealing Mithras' color with the ICVBC mobile lab in the museum. Heritage **2**, 2160–2170 (2019). https://doi.org/10.3390/heritage2030130
19. Merzbach, S., Hermann, M., Rump, M., Klein, R.: Learned fitting of spatially varying BRDFs. Comput. Graph. Forum (2019). https://cg.cs.uni-bonn.de/svbrdfs/
20. Merzbach, S., Weinmann, M., Klein, R.: High-quality multi-spectral reflectance acquisition with X-rite TAC7. In: Eurographics Workshop on Material Appearance Modeling, MAM@EGSR 2017. Eurographics Association (2017)
21. Müller, G., Tautges, J., Gress, A., Rump, M., Hermann, M., Lamy, F.: AxF - appearance exchange format. Technical report, X-Rite, Inc., 4300 44th St. SE, Grand Rapids, MI 49505, February 2020. https://www.xrite.com/-/media/xrite/files/whitepaper_pdfs/axf/axf_whitepaper_v18_en.pdf
22. Pfluger, D., Garmsen, L., Trumpy, G.: Modern methods for the visualization of lenticular film colors. Color Cult. Sci. J. **14**(1), 64–72 (2022)
23. Pliny the Elder: Natural History, Volumes I-X. Loeb, Cambridge (MA) (1938–1962)
24. Reuteler, J., Fornaro, P., Gschwind, R.: Digital Color Reconstruction of Lenticular Film: Project 'doLCE', June 2013
25. Series, B.: Parameter values for ultra-high definition television systems for production and international programme exchange. In: Proceedings of the ITU-T, Bt. 2020, pp. 1–7 (2012). https://glenwing.github.io/docs/ITU-R-BT.2020-1.pdf
26. Sinha, S.N., et al.: An Atlas-based approach for appearance-aware virtual 3D restoration and simulation of fading in fugitive textiles. In: Campana, S., et al. (eds.) Digital Heritage. The Eurographics Association (2025). https://doi.org/10.2312/dh.20253147
27. Sinha, S.N., Kuehn, P.J., Koppe, J., Kuijper, A., Weinmann, M.: Neural restoration of greening defects in historical autochrome photographs based on purely synthetic data (2025). https://arxiv.org/abs/2505.22291
28. Siozos, P., et al.: Light source classification and colour change modelling for understanding and predicting pigments discolouration. J. Phys. Photonics **7**, 035020 (2025). https://doi.org/10.1088/2515-7647/ade8c2
29. Streamlit Inc.: Streamlit (2019). https://streamlit.io. Accessed 01 Oct 2025
30. Taylor, H.A.: Harold Taylor slide collection. California Revealed (1896). https://californiarevealed.org/search?search_api_fulltext=&f%5B0%5D=search_page_series_title:Harold%20Taylor%20Slide%20Collection. Accessed 26 Nov 2024
31. Theophrastus: De lapidibus. Teubner, Stuttgart (1952)
32. Trumpy, G., Hardeberg, J.Y., George, S., Flueckiger, B.: A multispectral design for a new generation of film scanners. In: Optics for Arts, Architecture, and Archaeology VIII, Munich, Germany, July 2021, vol. 11784, pp. 138–146. SPIE (2021). https://doi.org/10.1117/12.2592655
33. Valero, E.M., Martínez-Domingo, M.A., López-Baldomero, A.B., López-Montes, A., Abad-Muñoz, D., Vílchez-Quero, J.L.: Unmixing and pigment identification using visible and short-wavelength infrared: reflectance vs logarithm reflectance hyperspaces. J. Cult. Herit. **64**, 290–300 (2023). https://doi.org/10.1016/j.culher.2023.10.016
34. Vitruvius, M.P.: De architectura libri decem. Teubner, Stuttgart (1991)
35. X-RITE: TAC7-Scanner (2018). http://web.archive.org/web/20180615015942/, https://www.xrite.com/categories/ap

36. X-RITE: Pantora material hub (2019). http://web.archive.org/web/20190424232441/, https://www.xrite.com/categories/appearance/pantora-software
37. Zhu, J., Park, T., Isola, P., Efros, A.A.: Unpaired image-to-image translation using cycle-consistent adversarial networks. In: ICCV, pp. 2242–2251 (2017)

Tools and Services for Colour Change Analysis and Sharing of Knowledge

Irina-Mihaela Ciortan[1], Bruno Fanini[2], Paul Julius Kühn[3], Marcello Massidda[2,4], Paschalis Panteleris[5], George Papadopoulos[5], Marios Pitikakis[6,7], Saptarshi Neil Sinha[3], Panagiotis Siozos[7], Sophia Sotiropoulou[7,8(✉)], Petros Ioannis Stavroulakis[7], and Giorgio Trumpy[1]

[1] ColourLab, Norwegian University of Science and Technology, Gjøvik, Norway
{irina-mihaela.ciortan,giorgio.trumpy}@ntnu.no
[2] Institute of Heritage Science, National Research Council, Florence & Rome, Italy
{bruno.fanini,marcello.massidda}@cnr.it
[3] Fraunhofer (IGD) Institute for Computer Graphics Research, Darmstadt, Hesse, Germany
{julius.kuehn,saptarshi.neil.sinha}@igd.fraunhofer.de
[4] Department of Antique Studies, Sapienza University of Rome, Rome, Italy
[5] Institute of Computer Science, Foundation for Research and Technology-Hellas (FORTH), Heraklion, Crete, Greece
{padeler,gpaps}@ics.forth.gr
[6] Department of Computer Science, University of Crete, Heraklion, Crete, Greece
pitikakis@csd.uoc.gr
[7] Foundation for Research and Technology-Hellas (FORTH), Institute of Electronic Structure and Laser, Heraklion, Crete, Greece
{psiozos,sophiaso,pstavroulakis}@iesl.forth.gr
[8] School of Applied Arts and Sustainable Design, Hellenic Open University, Patras, Greece
sophiaso@eap.gr

Abstract. This chapter introduces a suite of digital tools and systems that were developed within the PERCEIVE project to address challenges related to the conservation and communication of colour in cultural heritage collections. The authors, who developed these tools, describe the context and motivation behind, as well as their key features and functionalities. They also present demo use cases to showcase the performance of the tools. A key focus of the PERCEIVE project was the development of the Colour Knowledge Repository, which adheres to the FAIR principles for data sharing and reuse. The PERCEIVE platform offers the CH community centralised access to a variety of colour analysis tools, encouraging broader adoption and future integration. Tools such as MuLaX allow users to visualise and interact with information about the materials, techniques and preservation state of objects, enabling them to meaningfully discover scientific information. StyleShade3D assists experts in developing digital reconstructions of 3D objects, enabling them to apply different styles and shades to various parts of a 3D model using machine learning methods. The Light Damage Estimator provides digital simulations

I. M. Ciortan et al. (eds.), Chromatic Visions, LNCS 16450, pp. 57–96, 2026.
https://doi.org/10.1007/978-3-032-17700-1_2

and visualisations to predict colour changes in photosensitive materials exposed to light. The aim is to facilitate informed decision-making regarding exhibition lighting and long-term preservation strategies. Two tools developed in PERCEIVE focus specifically on autochromes. Deep Degreening uses AI restoration models based on networks that have been trained using synthetic image datasets simulating genuine greening defects to remove such defects from digitised autochromes. The Text2Autochromes uses generative AI to produce autochrome images. Finally, The Lenticular Colour Reconstruction Tool allows to recover the original colour information embedded in early lenticular films such as Kodacolor. All tools are now fully functional and have undergone testing to ensure stability and consistent performance.

Keywords: Digital tools · Visualisation · Prediction · Data-driven models · AI restoration models

1 Introduction

The PERCEIVE project (Perceptive Enhanced Realities of Coloured Collections through AI and Virtual Experiences) addresses a long-standing challenge in Cultural Heritage (CH) conservation and communication: fragility, loss, and misrepresentation of colour in artefacts.

To support conservation strategies for identifying, authenticating and preserving colour in heritage collections, PERCEIVE has focused in particular on developing digital tools that empower researchers and professionals to study, understand and interpret colour in a variety of heritage collections, as well as enabling the effective communication of scientific data. This provides a solid basis for any solutions aimed at reconstructing the original colours, thus reappropriating authenticity and the perception of the value of colour in its original context and concept.

In this chapter, the authors—the teams that conceived and collaboratively developed the PERCEIVE tools—describe the context and motivation behind the tools, as well as their key features, functionalities, use scenarios, typical workflows, and application contexts. By engaging with different subgroups of the heritage community, the developers aimed to address stakeholder expectations for improved access to heritage collections. Together with stakeholders, they co-created pilots that explored the opportunities offered by these digital technologies. These pilots are based on the five PERCEIVE scenarios, each of which represents and attracts a specific heritage subcommunity. This encourages participation in co-design and the provision of feedback, ensuring that the offered tools are aligned with sustainable values such as the communication of scientific knowledge, the enabling of better-informed decisions, and the increase of public engagement.

In the museum context, the preservation, authenticity and perception of colour form an interconnected loop. Scientific knowledge about colour lies at the core of this cycle, enabling the identification of original colours, guiding conservation strategies for preservation and informing debates about authenticity. Ultimately, this knowledge enhances perception and interpretation, broadening accessibility to these cultural heritage collections and strengthening public engagement with them.

Recognising the central role of scientific knowledge in managing, preserving their physical integrity, and communicating their cultural value, PERCEIVE prioritised **knowledge sharing**. To ensure long-term wide **access to scientific data** on the preservation or deterioration of colour in heritage objects, all primary data acquired for the development of digital tools, as well as the output data from these tools, services and prototypes, is made accessible for study and reuse via the **Colour Knowledge Repository**, described in Sect. 2.

Built according to FAIR principles, the **Knowledge Repository** enables advanced discovery and open access to data. Thanks to full documentation and interoperable file formats and information, this data can be reused by its creators and any other researcher. This data includes information on the state of preservation and susceptibility to deterioration of colours, as well as insights into their original appearance. However, datasets, technical images and 3D models acquired through advanced methodologies are often used exclusively by heritage scientists, conservators and data scientists, as they are the only ones who can understand and interpret them. PERCEIVE aimed to make the knowledge contained in this data accessible not only to scientists and experts, but also to non-experts and museum professionals, by designing tools that provide comprehensible visualisations of technical and analytical data.

The tools and services developed as part of the PERCEIVE project can be standalone, used independently by experienced researchers. However, PERCEIVE has created a platform that enables everyone in the CH community to access them easily. The **PERCEIVE platform** (see Sect. 3), provides a single entry point through which tools, services and knowledge are aggregated and made accessible to all. Furthermore, the platform could serve as an integration point for all colour analysis tools and services developed by the CH community in the future.

Tools such as **MuLaX**, detailed in Sect. 4, allow users to visualise and interact with information about the materials, techniques, and preservation state of objects, enabling them to discover scientific information in a meaningful way. For example, this takes the form of layers that users discover interactively on objects. The aim is to make this information accessible and usable directly, so that it can inform the decisions of conservators and enable the creative industries to develop public engagement applications.

StyleShade3D (see Sect. 5) is a tool designed to assist experts such as artists, designers, and curators in developing digital reconstructions of 3D objects. It enables users to easily apply different materials and visual styles to specific areas of the 3D model, showcasing potential restorations of objects such as classical statues. The tool aims to enhance the process of segment-wise rendering by enabling users to apply distinct styles and shading to different regions of a 3D model using learning-based methods. Generating a segmentation atlas—a

2D parameterised UV map over the 3D surface representing various segments of a 3D model—is required for the tool to function. StyleShade3D then leverages this semantic information, deep stylisation models, and shading models to enable region-specific stylisation of the surface materials of 3D models while preserving shading.

The **Light Damage Estimator**, described in Sect. 6, provides digital simulations and visualisations to show the effect that agents of deterioration have on objects over time. Designed as a tool to predict and visualise colour changes in photosensitive materials exposed to light, the Light Damage Estimator (LDE) aims to facilitate informed decision-making on exhibition lighting and long-term preservation strategies. It is primarily used in the preventive conservation of paintings and works on paper, though its framework can be transferred to other sensitive media, such as textiles and photographs. The LDE supports digital image inputs and provides visualisation outputs, offering interactive risk assessments. The tool integrates experimental pigment ageing data with the spectral power distribution of light sources to enable simulations of cumulative damage based on scenarios involving long-term exposure of objects to light.

Developed by the Lumière brothers in the early 1900s, autochromes represent a pioneering advancement in colour photography and occupy a vital place in the history of the visual arts. Each autochrome plate is a unique transparency image with delicate colour qualities. Their composition makes them fragile and sensitive to light, resulting in irreversible damage when they deteriorate. Two tools have been developed in PERCEIVE with a specific focus on autochromes.

The **Deep Degreening**, detailed in Sect. 7, is a tool that can be used to restore digital images of autochrome plates displaying greening defects. The de-greening tool uses an AI service to remove the greening defects from the input image. AI restoration models have been based on networks trained with synthetic image datasets simulating genuine greening defects, which typically appear as small, point-like stains or broader leakage-style areas.

While the Deep Degreening tool uses AI restoration models that enable the removal of greening defects from digitised autochromes by creating synthetic greening defects using image processing techniques, **Text2Autochromes** (see Sect. 8) uses generative AI to generate autochrome images from text, including images with greening defects. The tools developed within PERCEIVE project framework are now fully functional and have undergone testing to ensure their stability and consistent performance. At this stage, they are configured to operate with specific case studies that serve as demonstrators, showcasing their functionalities and potential in realistic scenarios. These demonstrations also provide the basis for evaluation by end users, who assess the tools' usability, performance, and relevance to their professional needs. Following this initial evaluation phase and the systematic collection of user feedback, the tools will be further refined and upgraded to support open use. This next stage will allow users to apply the tools flexibly to their own, user-defined case studies, broadening their applicability and impact beyond the demonstration context.

2 Colour Knowledge Repository

2.1 General Description and Scope

The PERCEIVE Colour Knowledge Repository is a data management system developed to collect, organise, and openly share colour-related resources in cultural heritage artworks.

Encompassing artifacts ranging from ancient polychrome sculptures and delicate painted surfaces to dyed textiles and historical film, the repository facilitates the sharing of documentation and simulation data concerning colour changes, as well as datasets supporting the assessment of deterioration risks. In addition, it provides a platform for sharing protocols and workflows dedicated to the preservation of coloured collections and the mitigation of further deterioration. Designed as a collaborative research environment, the Repository brings together a diverse community of stakeholders—including colour experts, scientists, museum curators, conservators, exhibition designers, and education and communication professionals—working across physical and digital cultural heritage domains.

The homepage of the Colour Knowledge Repository is accessible at: https://perceive-data.iesl.forth.gr (Fig. 1).

2.2 User Needs and Technical Approach

Identified User Needs. The study and characterisation of the materials and techniques used in the creation of artworks, the analysis and interpretation of colour changes and alterations, the implementation of preventive conservation strategies, and the careful and appropriate display of coloured collections require interdisciplinary collaboration. Such collaboration involves a wide range of professionals—researchers, scholars, curators, conservators, lighting designers, and communication and cultural industry professionals- whose expertise converges around the colour of artworks. Effective cooperation relies on mutual understanding, the sharing of knowledge, continuous updating of data, and the enrichment of knowledge through systematic monitoring across the lifecycle of cultural objects.

The Repository has been designed to address these needs by providing a shared, openly accessible environment [10] where users can find, analyze, interpret, reuse, and update information. It facilitates access to publications, reports, and interpretative studies directly connected to raw or processed datasets. It provides access to a wide variety of resources in various formats and containing different types of information, including historical and artistic context, analytical data, processing and simulations, and digital reconstructions. These interconnected resources are searchable through a unified interface, enhancing discoverability and use.

Our goal is to establish a data management infrastructure that supports the full lifecycle of data—from acquisition to processing, interpretation, and beyond. Primary data, acquired through a repertoire of analytical techniques applied to

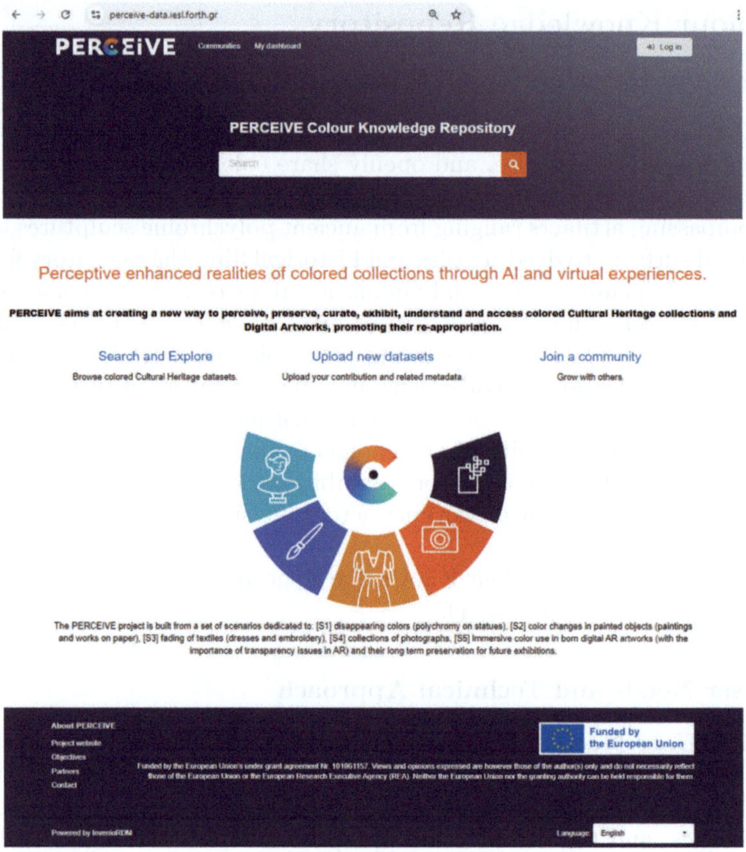

Fig. 1. The home page of the Colour Knowledge Repository

physical heritage objects, captures complementary aspects of material identity, including chemical composition of pigments and colourants, physical properties of colour and texture (and their variability under different lighting conditions), the condition of coloured surfaces, and the causes and chemistry of colour change.

However, primary data are rarely self-explanatory, advanced processing techniques are applied to extract meaningful information, which is often hidden in the raw data. This processed data may then be combined to construct simulation models that enable reconstructions of previous states or predictions of future changes in the appearance of coloured objects.

Given that such datasets and resources are obtained at different times, by different contributors, and sometimes iteratively, meaningful reuse requires rigorous documentation, complete metadata, and semantic interlinking. The Repository, therefore, organises resources within a structured, interoperable framework that reflects the entire data lifecycle, ensuring that they remain comprehensible, reusable, and valuable to the interdisciplinary community.

Technical Architecture and Frameworks Used. The Colour Knowledge Repository is based on InvenioRDM, an open-source framework developed by CERN for building large-scale digital repositories. Even though Zenodo [25] has broad visibility in the Open Science community, making it an excellent choice for sharing scientific outputs at no cost, its fixed metadata structure and lack of customisation can limit its functionality. Compared to Zenodo, deploying a self-hosted InvenioRDM instance offers significant advantages for tailored Research Data Management (RDM). It allows for richer metadata, semantic annotation, integration with workflows, fine-grained control, and alignment with FAIR principles [3,26].

Content and Storage. The repository hosts diverse resources on a wide range of topics, interests, and themes relevant to colour knowledge and preservation of colour in art and heritage objects. This includes 3D models, scientific datasets, high-resolution scientific images, visuals and graphics (charts, diagrams, maps, visual informative material), educational material guidelines and tutorials, textual resources (literature, research papers, historical text resources, scientific literature, technical reports), protocols, workflows, reference materials and other resources as glossaries, common vocabularies on colour related topics.

Even though it is possible to store datasets in the Colour Knowledge Repository, we decided to follow a more decentralised storage model. Instead of storing both data and metadata in a central repository, we store mostly metadata and links to the data. The actual data can remain at the contributor's premises or other data storage solutions across multiple locations. To prevent data consolidation, we established a Federated Cloud file-sharing system using NextCloud servers that aims to support collaboration, streamline data exchange, and offer a unified view.

This overall approach provides more flexibility, robust data management capabilities and control over the data/metadata, customisation options, integration with other research tools, and ensures compliance with legal, ethical, and security requirements.

Metadata and Documentation. The repository is equipped with an advanced metadata model that allows for rich description and categorisation of data. This includes detailed information about data acquisition methods, processing, and analytical techniques used. The metadata model captures all scenarios and aspects of colour addressed in PERCEIVE, ranging from meaning, appearance, authenticity, care, integrity, loss, change, damage, fading, measurement, rendering, reconstruction, and prediction. The use, interpretation and transformation of complex scientific data into knowledge is largely dependent on the relevance and completeness of metadata that documents the creation or acquisition process and describes the outputs of subsequent processing and analysis.

To ensure consistency across all data collected and uploaded to the Repository, and to ensure interoperability with other documentation systems, we adopted a well-established Heritage Science metadata framework as a reference.

More specifically, the PERCEIVE metadata model was developed in line with the concepts and principles of the European Research Infrastructure for Heritage Science (E-RIHS) approach [14], as it offers a broad and inclusive perspective on the documentation of heritage science data, encompassing a wide variety of research resources and applications. The E-RIHS model is structured in modular components that can be linked together to comprehensively document the processes of data creation.

The Colour Knowledge Repository is organised into a hierarchy of over 60 resource types that reflect different techniques, dataset categories, and data formats. The repository can store and search for resources like: Scientific datasets (research data, spectra, colour measurements), Literature textual resources (research papers, historical text, scientific literature, technical reports), Historical audiovisual archival documents (photographs, films, audios), 3D models of coloured heritage & art objects, Technical/Scientific images, Visuals and Graphics (charts, diagrams, maps, visual informative material), Guidelines and tutorials, Training material/educational resources, Standards, Protocols, Workflows, Research tools, Reference materials etc.

The metadata fields were extended in a structured way to capture detailed information on data acquisition and production methods, including processing steps, instrumentation, and experimental conditions or parameters. To support this, controlled vocabularies were established, and multiple metadata fields are constrained to values drawn from custom predefined vocabularies, ensuring consistent documentation of the research activities and the analytical or processing techniques applied. In addition, standard metadata terms and sub-models were adopted to record the methodology, the tools used (hardware or software), the experimental setup and protocols, as well as the nature and content of the resulting data. Other custom metadata fields were introduced, divided into three subcategories: *Research focus*, *Research activity* and *Supplementary information on Object Examination*. Furthermore, dedicated fields were added to document relations among different resources in a structured manner.

2.3 Key Features and Functionalities

The repository is based on an open architecture and well-established standards and practices, with a scalable infrastructure capable of supporting a growing volume of content and users. It is designed as a dynamic and evolving resource, capable of accommodating an expanding user community and an ever-increasing body of knowledge on colour. Its structure is aligned with the FAIR principles, ensuring that data is Findable, Accessible, Interoperable, and Reusable.

The main repository functionalities include the following:

- *Uploading and Managing Data*: Users can upload datasets along with a rich set of associated metadata. The system supports metadata-only records that link to external data sources or files stored directly in the repository. Flexible visibility options allow users to make data and metadata public, keep files restricted while keeping metadata public, restrict both files and metadata, or apply an embargo period. Users can also choose the license under which a dataset is shared, such as a Creative Commons license, with the option to add a custom license if needed. Legal and ethical considerations are prioritised to ensure responsible knowledge management.

- *Communities and Collaboration*: Users can create and manage their own communities within the repository (either public or restricted), and invite other registered users to collaborate and share data within groups. Currently, dedicated communities have been created to exchange data and knowledge related to colour in PERCEIVE scenarios (i.e. colour loss in originally polychrome ancient sculptures, colour changes in paintings and light-sensitive-coloured objects, colour fading of dyed historical garments and textiles, and colour in historical photography and films).
- *User Management and Role-Based Access Control*: The repository offers personalised accounts where users can manage their data, set privacy settings, and customise their interaction with the system. Users can currently connect with their ORCID credentials, but other options are also possible (e.g. OAuth, SAML, Single Sign-on). Different roles are available for community members, ranging from read-only access to full administrative rights.
- *Security and Compliance*: The system adheres to stringent security protocols and compliance standards, including GDPR, to ensure data protection and privacy.
- *Searching and Accessing Data*: Advanced search and filtering capabilities help users to discover datasets efficiently. Users can search by keywords or phrases, metadata fields, Boolean operators, and use filters to refine results. Faceted navigation allows users to browse records by selecting categories or facets such as resource type, techniques and more. Users can view or edit records that they own or manage (e.g. community curators). Several data types and file formats are also available for online preview before downloading them.
- *Exporting and Integrating Data*: Data and metadata can be exported in various formats that correspond to widely used metadata standards such as JSON, DataCite, Dublin Core, BibTeX, MARCXML, DCAT-AP, etc. The system supports integration via a REST API, enabling automation and interoperability with external platforms. Developers can programmatically search and retrieve records or harvest metadata using the OAI-PMH protocol.

2.4 Impact, Outreach and Future Developments

The PERCEIVE Colour Knowledge Repository is a domain-specific repository for Heritage Science that goes beyond serving as a storage solution. It is a dynamic platform that strengthens the research landscape by enabling efficient data management, promoting interdisciplinary collaboration, and fostering innovation in heritage science and related fields. Designed as an open, inclusive, and scalable knowledge resource, the PERCEIVE Colour Knowledge Repository addresses both current and future stakeholder needs on colour, as defined in the PERCEIVE framework. It encompasses a broad spectrum of topics, interests, and themes, and aims to enhance the visibility and impact of research in colour knowledge. It supports open science initiatives by making data accessible to a broader audience, thereby fostering collaboration and innovation.

The PERCEIVE Colour Data Repository can transform the way colour in Digital Cultural Heritage and Digital Art is communicated, making the results and the collected datasets accessible for immediate use and further development, under the proper IPR restrictions and regulations. Once the repository is sufficiently populated, its use will allow wider access to visitors and the ability to explore and discover coloured collections and their restoration, improving their understanding and increasing both visitors' and stakeholders' engagement and participation. It also supports the principles of open science, where data and methodologies are shared transparently, which improves the reproducibility of research findings and builds trust within the scientific community and the public, and advances the overall dissemination and utilisation of knowledge.

Future developments include further integration with the PERCEIVE platform and enhancements to user interaction, data visualisation, and analytical capabilities. In addition, we plan to extend the accessibility and incorporate mappings to the standard CIDOC-CRM ontology or other semantic annotations and knowledge services for more intelligent resource discovery and reuse. Scenario-specific knowledge graphs can allow data to be enriched and its context enhanced, enabling advanced searching and visualisation services.

3 Perceive Platform

3.1 General Description and Scope

The tools and services developed under the scope of the PERCEIVE project can be used "standalone" by experienced researchers. However enabling easy access to everyone in the CH community is a difficult task. The PERCEIVE project aspires to solve this with the **Perceive Platform**, a central point where tools services and knowledge is aggregated and made accessible to all. Furthermore the platform can be an integration point for all colour analysis tools and services developed by the CH community in the future. There are three main components in the perceive platform: The Backend or PerceiveAPI, the Frontend and the Workers.

The **Backend** (also referred to as the PerceiveAPI) constitutes the central integration layer of the PERCEIVE platform. It is the "nervous system" that orchestrates communication among tools, services, and the user-facing Frontend. While individual tools (e.g., light damage estimation, De-greening, LCR, etc.) deliver specific functionality, the PERCEIVE API ensures that they interoperate seamlessly as part of a unified infrastructure. Developed using the **.NET Aspire** microservices architecture, it enables scalability, modularity, and resilience. Each microservice operates independently in its own containerised environment, communicating through RESTful endpoints and, where efficiency demands, through gRPC channels. The Backend also acts as a broker for job orders submitted from third parties (I.e users using the Frontend): Jobs are delegated to Workers through a message queue service and job results are gathered logged, stored and made available by the Perceive REST API.

The **Worker** is a containerised application that wraps the functionality of specialised Perceive tools. A worker can run in the same cluster as the backend or be hosted on a completely different network or even run locally on a researchers workstation. On startup the worker connects to the backend (PerceiveAPI) and registers it self to execute job orders through the message queue service. Since a worker is required to run jobs that might need specialised hardware (i.e. GPUs) it also declares its capabilities to the backend service. This way it only receives jobs that it has the resources to complete. Workers can be added or removed from the platform seamlessly, enabling scalability and robustness.

Finally the **Frontend** serves as a lightweight Web user interface for the PerceiveAPI (see Fig. 2). It enables users to submit jobs and see results using any web-enabled (PC, tablet, phone) device without any special hardware requirements.

The design prioritises flexibility: tools can be integrated, extended, or replaced without disrupting the rest of the system.

The PerceiveAPI is not just a technical necessity but a strategic enabler of PERCEIVE's long-term objectives:

- **Scientific integration:** aligning heterogeneous datasets and analysis outputs across tools
- **User empowerment:** offering consistent, secure access to tools for conservators, curators, and creative industries
- **Knowledge sharing:** ensuring that workflows, metadata, and results flow into the Colour Knowledge Repository for reuse and reproducibility.

By acting as the backbone of the platform, the API guarantees that the principles of CARE and FAIR data, authenticity, accessibility, reusability, are embedded into every interaction.

3.2 User Requirements and Technical Approach

In the progress of the PERCEIVE project, several common needs across stakeholders where identified:

- **Unified Access:** Experts and museum professionals require a single entry point to interact with multiple tools without having to manage separate logins or configurations
- **Interoperability:** Researchers want to link simulation data (e.g. light damage estimation) with analytical layers (e.g. hyperspectral images) through consistent APIs
- **Security and Trust:** Institutions demand robust authentication, role-based authorisation, and compliance with GDPR
- **Extensibility:** Developers require integration guidelines so that new tools can be added to the ecosystem with minimal friction.

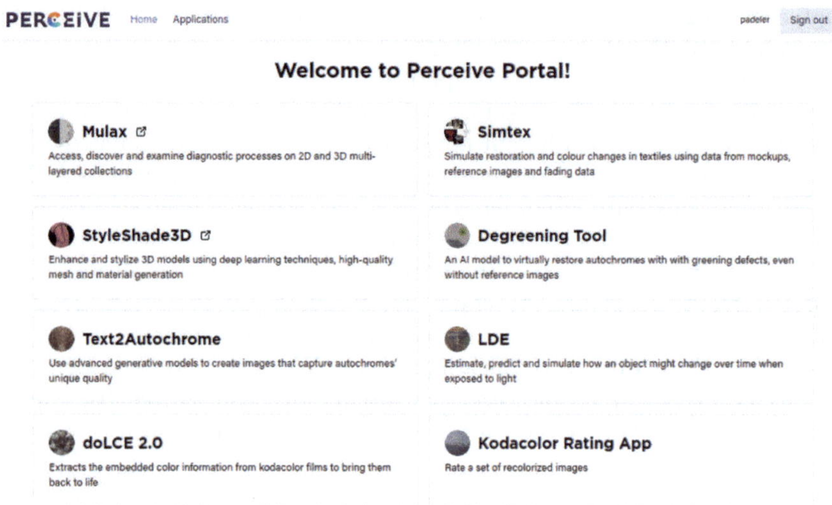

Fig. 2. The home page of the Frontend. All tools are readily available through a simple to use web UI.

The Perceive Platform addresses these needs through the following design principles:

- **Microservices architecture:** each functional unit (data service, identity service, metadata service, orchestrator) runs independently and can scale horizontally
- **Backend-for-Frontend (BFF):** a dedicated layer mediates between frontend interfaces and backend services, simplifying client complexity
- **Containerisation with Docker:** ensures consistency across development, testing, and production environments
- **gRPC integration:** complements REST by supporting high-performance communication between tools
- **Message Queue:** Orders are delegated through a Message Queue service (RabbitMQ). This allows specialised workers to be added or removed with minimal impact to the platform operation adding to the scalability and robustness of the architecture (Fig. 3).

3.3 Key Features and Functionalities

The Perceive Platform supports a range of features central to the PERCEIVE system:

- **Authentication & Authorization:** Implemented via **OAuth 2.0**. Users authenticate through an Authorization Server that issues short-lived JWT access tokens and refresh tokens. Role-based access control ensures tailored

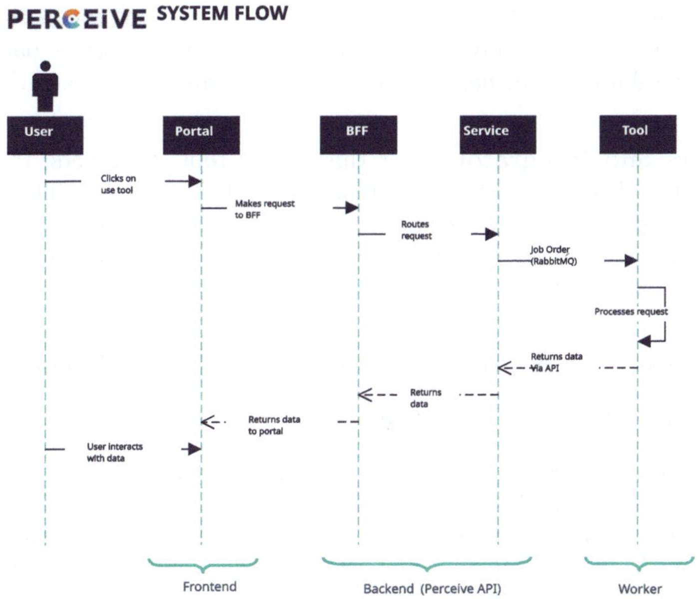

Fig. 3. Information flow in the Perceive API for a typical used interaction through the frontend. The User submits a job request which is delegated through the BFF to the corresponding service and finally executed by a worker. All results are logged, stored and made available to the user.

permissions (Visitor, Expert, Admin, Developer). Additionally OAuth 2.0 enables **third party authentication** (I.e via ORCID) which allows for a seamless integration with 3rd party tools like the InvenioRDM of the Colour Knowledge Repository.
- **Data Management:** A dedicated Data Service manages storage and retrieval, integrated with **Nextcloud federated file systems**, allowing decentralised storage with centralised access.
- **Tool Orchestration:** The Orchestrator coordinates multi-step workflows, routing requests to appropriate tools and managing intermediate results.
- **Error Handling and Logging:** All endpoints return standardised error codes and logs feed into central monitoring, ensuring traceability and system resilience.

3.4 Use Scenarios and User Profiles

Experts and Conservators: An expert analysing the fading of a painting request access to relevant tools and data via the API, run a colour change simulation through the Light Damage Estimator tool, and automatically store both results and metadata in the Repository. The API manages authentication, orchestrates tool invocation, and ensures results are linked.

Museum Visitors and Educators: In an educational scenario, a visitor interacts with a WebXR demo powered by MuLaX. The API provides backend support: fetching datasets, authorising access, and streaming reconstructions.

Developers and Integrators: For third-party tool developers, the Perceive API provides clear documentation (Swagger interface, endpoint definitions, templates) and standard integration pathways. A new colourisation tool, for instance, can be registered, tested locally via Docker, and orchestrated into workflows through the API.

4 MuLaX and Analytical Data Visualisation Tool

4.1 General Description and Scope

"MuLaX" (Multi-Layer eXtensible eXplorer) is a Web3D/WebXR tool [7] built on the open-source ATON framework [5] especially for the PERCEIVE project. It allows users to access, discover and examine diagnostic and analytical data available for 2D and 3D multi-layered artefact collections namely paintings, textiles, sculptures and frescoes. The main focus of MuLaX is to enable the interactive discovery of invisible (or lost) information made accessible through diagnostic processes. It offers an increased readability of the artwork while exploring new ways to access and examine multi-dimensional items, directly in a web browser.

4.2 User Requirements and Technical Approach

Identified User Needs. The MuLaX tool has been designed to address the needs of two main target groups.

The first target group comprises experts, including conservation scientists, academic researchers, and museum professionals. Their core need arises from the fragmented nature of diagnostic practices: each analytical method (e.g. multispectral or hyperspectral imaging, spectroscopy, microscopy, graphical plots) produces outputs that differ in scale, format, and nature. This diversity makes collaboration difficult, as there is no unified environment in which to visualise all results together. Experts therefore need a web-based 3D platform that enables them to align heterogeneous data on the same digital replica of an artefact, switch seamlessly between datasets, and refer to the same point, area, or layer, facilitating effective communication and collaborative interpretation.

The second target group is the general public, including museum visitors, students, and schools. Their needs are centered on outreach and communication. For them, it is not enough to see visual reconstructions of cultural heritage artefacts; it is essential to understand the scientific evidence that supports these reconstructions. MuLaX addresses this by enabling the rapid creation of tailored exhibits and interactive demonstrations, where curated diagnostic datasets can be explored directly on 3D artefacts. In this way, non-experts can engage with the data behind reconstructions, gaining awareness of heritage science while experiencing cultural artefacts in an accessible, interactive experience.

Technical Architecture and Frameworks Used. ATON is an open-source framework [5] developed by CNR ISPC, to present and interact with 3D models and scenes online, offering several functionalities targeting Heritage Science communities. The framework provides an adaptive presentation layer, with the creation of Web3D applications with interactive, liquid 3D visualisation. Such Web3D applications can be consumed and accessed on every device, ranging from smartphones, museum kiosks, workstations, up to immersive XR devices, without any installation required for final users. A few components and features of ATON were specifically used targeting MuLaX tool:

- Collections: crucial to manage 2D/3D items and other data using ad-hoc routines, also providing smooth integration of data layer with external clouds
- Dynamic materials: using physically-based rendering (PBR) to simulate specific surface properties, as well as custom effects to implement local discovery routines of multiple scientific layers
- Semantic annotations: crucial to enrich and annotate items with diagnostic processes and other information interactively queried by users
- Spatial UIs: dedicated components to easily build interfaces targeting immersive VR modes or AR, providing new ways to inspect multi-layered items
- Built-in authentication: component to control execution of specific tasks (e.g.: adding or modifying a POI can be performed only by a restricted team)

The MuLaX tool was developed on top of the "plug and play" web-app architecture [6] offered by the ATON framework. This allows the tool to be coherent and well integrated with the underneath ecosystem, including routines to access centralised resources (e.g. 3D collections, etc.), UI templates, as well as built-in routines to facilitate application setup and deployment. The Web3D tool supports the inspection of multiple 2D or 3D items: such resources can be hosted on the same server node, dedicated clouds or remote storage (Fig. 4). In order to maximise integration, involved items are placed into the main collections folder of the ATON framework and accessed by dedicated routines. Alternatively, external data services or public folders (accessible via web services) can be used in order to fetch required assets.

Interoperability with Standards. The adoption of ATON framework provides robust interoperability through the use of international ISO standards like glTF format [19] designed by Khronos group. Such open standards guarantee long-term preservation, reuse and extensibility. glTF is in fact a royalty-free specification that bridges the gap between 3D content-creation and modern Web3D applications by offering an efficient, extensible, interoperable format for the transmission and loading of 3D content. Within MuLaX, this format can be used to carry out additional information (in the form of maps/textures) to simulate specific materials and make them react to light setups (punctual/directional lights and/or image-based lighting). This can be done either via built-in glTF Physically-based Rendering (PBR) model, or via custom shaders that calculate punctual response to incident light rays.

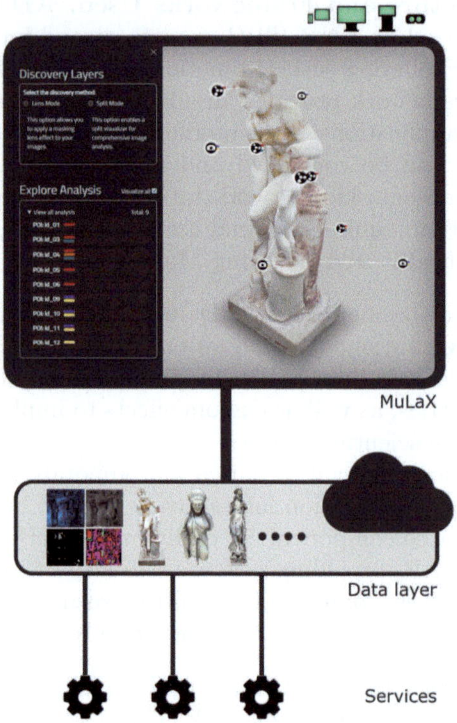

Fig. 4. Simplified overview of the architecture: MuLaX Web3D tool (top) has access to data layer (cloud collection - middle) where also processing services may have access to operate (bottom).

4.3 Key Features and Functionalities

The features and functionalities of the MuLaX tool can be explored in two operating modes, depending on whether the data being investigated covers the entire surface or a specific point on the object (i.e. layer and point-type data overlays).

Layer Investigation Mode. A prominent feature of MuLaX tool is the multi-layer discovery: the original lens metaphor [24], was inherited from previous CNR ISPC projects based on ATON, operating in multi-temporal contexts [4] and in the diagnostic field [8,18].

For data collected over the entire object (layer data), MuLaX can load it underneath the natural colour layer - typically visible information. It can then be inspected interactively using either the lens or the split view tool. The Lens Tool allows to interactively and locally discover additional data beneath the colour layer or visible information. As the user moves the tool over the surface of the 3D or 2D object, the required layer is revealed within the lens radius. This allows the user to 'explore' the surface, focusing attention on the discovered area.

Alternatively, another exploration mode named split view tool, can be used to progressively reveal the hidden bottom layer in the X, Y, or Z axes. This mode is controlled using the mouse (or a VR controller) to define the split boundary.

To depict these two modes, in Fig. 5 we show an example of the Split View and lens tool for the Scream (1910?) painting by Munch use case. The same underlying data is used in both cases, namely a hyperspectral image slice at a wavelength of 500 nm. Note that the for paintings the Split View does not operate along the Z direction since that is the direction normal to the painting.

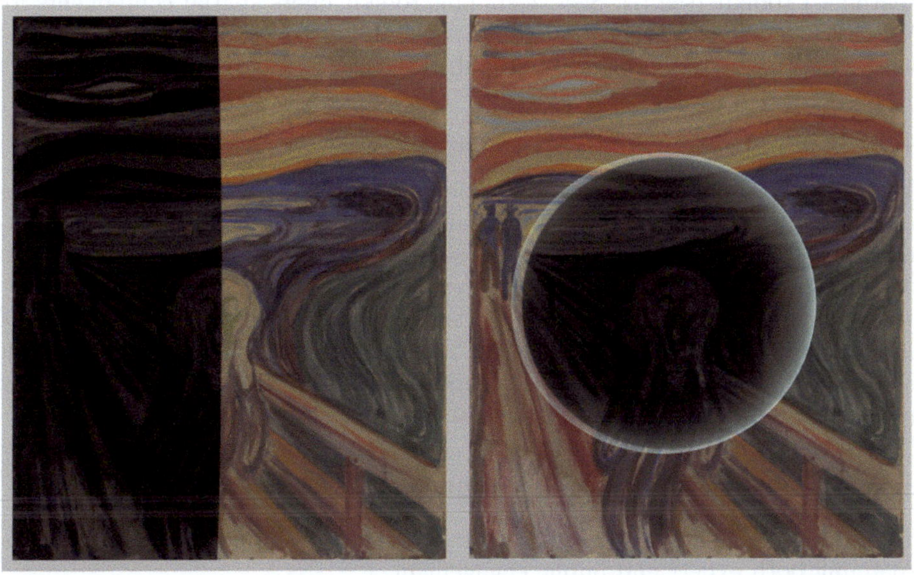

Fig. 5. A sample layer discovery in MuLaX, applied to a 2D object to examine analytical data collected on the painting 'The Scream' (1910?) by E. Munch, depicting the Slit View operating along the X-axis (left) and the lens tool (right). The underlying layer in both cases is a hyperspectral image slice at a selected wavelength of 500 nm.

Point Data Mode. This mode enables users to examine data collected from a specific area or region of the object point by point. This data can be in the form of a graph (e.g. XRF, FORS, Raman spectra) or an image (e.g. a UV micro or macro photograph).

Annotations and Masks. MuLaX supports the creation of POIs (points-of-interest) to enrich the item punctually with associated data, as introduced in previous subsection. These POIs can be classified with categories or techniques, providing live filtering in MuLaX (see Fig. 6, A). Masks on the other hand, allows to define more complex, image-based semantic areas (see Fig. 6, C), furthermore offering in perspective selective style transfers or selective material parameters control on both 2D or 3D items. Masks can be fetched from other services, cloud storage or interactively generated by users via semantic painting (see [7], Sect. 4.2).

XR Inspection. The tool offers the possibility (via WebXR API) to present and inspect a given item in immersive VR, or mixed/augmented reality modes. This offers a way to inspect full-scale digital replicas through a spatial exploration. The tool also extends base ATON spatial UI components to provide a dedicated spatial UI toolkit (see [7], Sect. 4.4). This toolkit offers spatial components for VR or AR sessions to inspect and discover the item (see Fig. 6, D and E) enriched with POIs and masks previously described.

4.4 Integration with Other Tools or Platforms

The framework can be deployed in different modes thanks to a micro-services architecture based on Node.js [6]. This also offers a REST API that allows external tools or platforms to operate on selected resources and web-applications deployed. The data layer integration previously introduced (and detailed in [7]), allows MuLaX to fetch resources (3D models, layers, masks, etc.) from cloud storage, where for instance other PERCEIVE services are operating. To provide an example, data layers used in discovery tasks in MuLaX could be the result of previous computations performed by other services.

4.5 Use Scenarios and User Profiles

The MuLaX tool's structure has been designed to address two main use scenarios, spanning both expert and public engagement.

Experts. For experts—such as academic researchers, heritage scientists, and museum professionals—MuLaX offers a unified platform to annotate, compare, and visualise diagnostic data directly on 3D models of cultural artefacts. The typical workflow involves performing analyses (e.g., spectroscopy, multispectral imaging, microscopy), uploading the resulting data into MuLaX, and aligning them with the 3D virtual replica of the analyzed object [9]. Researchers can then enrich the dataset with descriptive images and structured metadata. Once the 3D dataset is configured within the platform, users are able to collaboratively explore the multi-layered data across desktop, mobile, or XR interfaces.

Public. For the general public (or visitors interested in heritage science, including schools and students), the same platform becomes a medium for interactive demonstration: curated diagnostic datasets, visual overlays, and descriptive annotations can be presented as guided digital exhibits. The workflow here

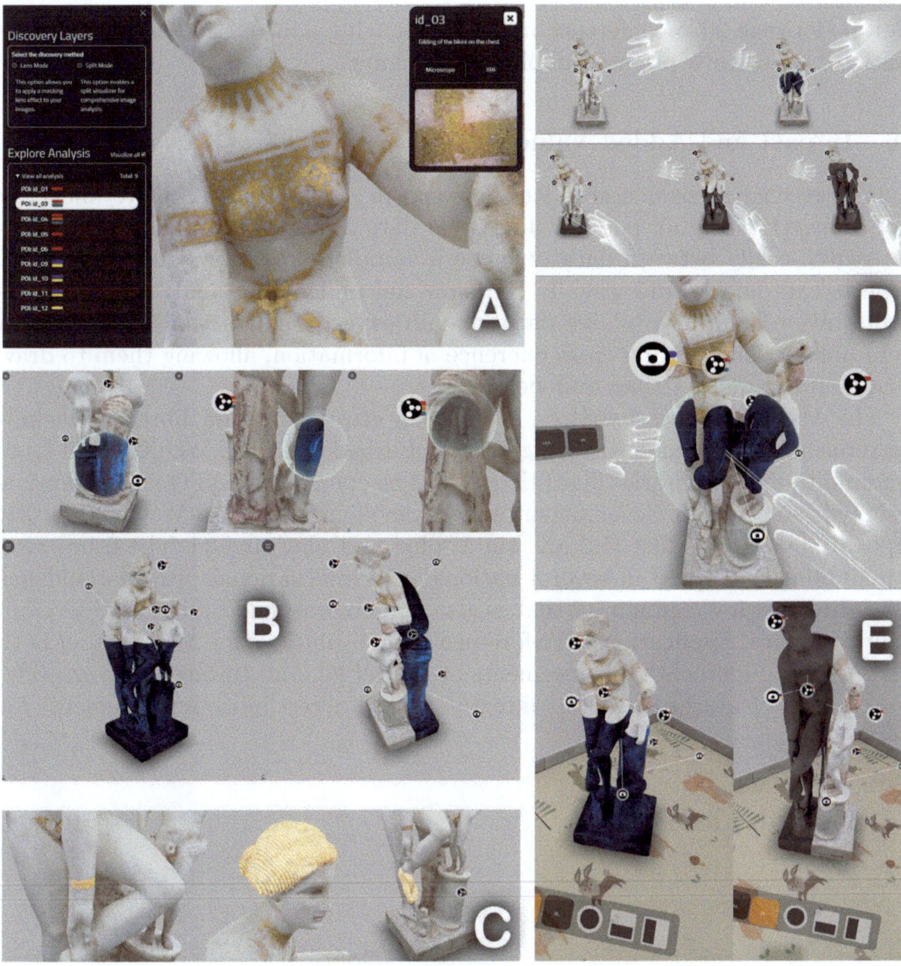

Fig. 6. A) Sample filtering and query of POIs on top of the item (bikini venus); B) Discovery layers (in this case UVL and VIL); C) Semantic masks; D) Immersive VR inspection and E) AR inspection.

focuses on the rapid assembly of content, designing a tailored narrative for the target audience, and deploying the demo for access via desktop, mobile, or immersive VR/AR environments. In this way, MuLaX serves not only as an analytical backend but also as a versatile exploratory tool, bridging rigorous scientific investigation with accessible public outreach.

As demonstration cases, MuLaX was applied to The Scream (1910?), showcasing the visualisation of diagnostic analysis on paintings represented as two-dimensional artefacts, and to the Venus Anadyomene, within the scenario of sculptural artefacts. Both case studies were presented as pilot exhibits at the

Digital Heritage 2025 conference in Siena, engaging a mixed audience of experts and heritage-science–oriented public participants.

4.6 Impact and Added Value

Mulax presents a considerable step forward in enabling an intuitive and easy way of investigating all possible analytical data available for an artwork. Users can switch easily between data types be it complete layers or point/areal data. Experts can quickly and visually compare and understand analytical measurements taken from all available data modalities: a paradigm which allows for a more holistic analysis and cross reference of information, allowing them to draw insights much quicker than ever before.

The MuLaX tool is based on large open ecosystems and libraries, including robust Web3D standards like glTF, established APIs like WebXR, and large community-driven libraries like THREE.js. The open-source ATON framework is designed to accelerate development and deployment of vertical Web3D/WebXR applications or tools like MuLaX, via a "plug and play" model. The framework is maintained, adopted in several national and international projects, and has a steadily growing community that is also gathered around design and development of vertical applications. We foreseen large evolutions for MuLaX tool both inherited by underneath advancements of ATON [6], and through new projects, bringing new features as well as integration and accessibility improvements for targeted users.

5 StyleShade3D

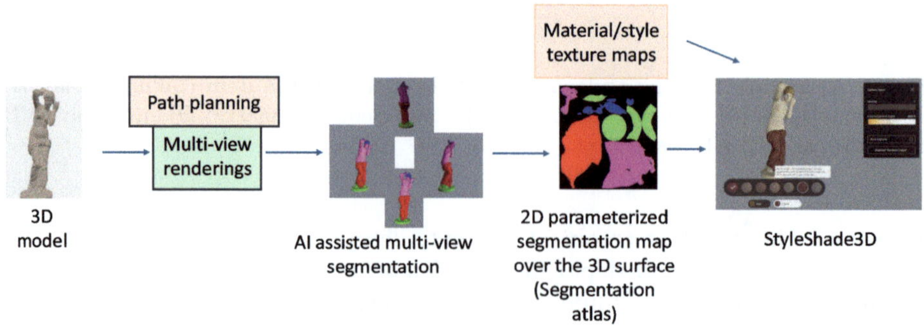

Fig. 7. Overview of the workflow for stylising and shading different segments of the 3D model and finally visualising using our StyleShade3D tool.

5.1 General Description and Scope

StyleShade3D is a tool that enhances the process of segment-wise rendering by allowing users to apply distinct styles and shading to different regions of a 3D model utilising learning-based methods. It is built on WebGL, a JavaScript API for rendering interactive 2D and 3D graphics in a web browser. The tool utilises a 2D parameterised segmentation map (segmentation atlas) [21] over the 3D surface, which is created through a combination of deep learning-based methods and traditional 3D mesh processing techniques to select the different segments.

The workflow (see Fig. 7) begins with generating a segmentation atlas, a 2D parameterised uv-map over the 3D surface that represents various segments of a 3D model. This is achieved by rendering multi-view images of the model using camera animations to ensure comprehensive coverage. We use an approach inspired by path planning approaches used in robotic applications to identify minimum set of views necessary to accurately represent the 3D surface.

Once the segmentation atlas is generated from the backend service (the tool for this backend service is ongoing work), StyleShade3D leverages this semantic information, deep stylisation models, and shading models to enable region-specific stylisation and shading. This allows artists, designers and curators to easily apply different materials and visual styles to specific areas of the 3D model to showcase potential restoration of the 3D objects like statues.

5.2 User Requirements and Technical Approach

Creating mockups of pigments based on scientific analysis is a common practice in conservation science. Additionally, photographic documentation and other artworks provide representations of the colours found in various 3D cultural heritage items. As a result, multiple colour simulations of the original 3D objects can be generated to explore potential hypotheses about their reconstructed colours. However, the criteria for segmentation can change with new evidence, making it challenging to train an AI model that works effectively across all 3D models.

To address these challenges, our workflow enables faster segmentation of regions using AI with user guidance. We focus on the image space, where segmentation algorithms are generally more advanced and accurate compared to 3D. This approach allows us to integrate high-quality scans of 3D objects, such as those obtained from laser scanners, with deep learning techniques like image segmentation, style transfer, material parameter estimation, and inpainting. This combination enhances the accuracy and efficiency of our segmentation process, ultimately contributing to efficient conservation outcomes.

5.3 Key Features and Functionalities

The backend service provides functionality to create various segments of a statue based on user input. Users can choose different materials for each segment, allowing for simulations that showcase potential colour variations. Additionally, the tool enables users to adjust the colour temperature of the light, enhancing the visualisation of the segments. It also presents detailed descriptions of the material properties for each segment, facilitating a better understanding of the choices available.

5.4 User Scenarios

In this user scenario, a conservation scientist conducts a scientific analysis of a 3D object, such as a statue. Based on this analysis, the scientist segments an image of the statue using our AI service and collaborates with technical experts, uploading the image to a data server. The scientist can also upload style images and material properties obtained from devices like Total Appearance Capture (TAC7).

Technical experts then use the segmentation image, guided by the scientist, to quickly generate a segmentation atlas. The tool allows for the application of different styles and materials to various parts of the statue. The final results are shared via a web link, enabling the conservation scientist to review all options and select the most plausible reconstructions.

5.5 Impact and Added Value

The AI-driven segmentation tool greatly improves how conservation scientists work on restoring objects like statues. It allows them to quickly create a segmentation atlas from uploaded images, saving time and letting them focus on important tasks. By collaborating with technical experts, they can achieve accurate reconstructions. The tool also offers the flexibility to apply different styles and materials, helping them to understand different potential restorations. With access to detailed information about materials, scientists can make better choices that maintain the original look of the statue. Sharing the final results through a web link helps professionals work together and learn from each other, ultimately enhancing the quality of restoration efforts and helping to protect cultural heritage.

5.6 Sustainability and Future Developments

The use of this AI-driven segmentation tool promotes sustainability in conservation by optimising resource utilisation. By streamlining the restoration process, the tool reduces the need for excessive materials and resources, leading to more environmentally friendly practices.

As the tool evolves, future developments aim to further enhance its capabilities. The backend service will be integrated as an additional tool, minimising the reliance on technical experts and making the technology more accessible to conservation scientists. Additionally, the tool will support the export and import of reconstructions, facilitating easier sharing and collaboration across projects.

6 Light Damage Estimator

6.1 General Description and Scope

The Light Damage Estimator (LDE) is a web-based decision-support tool developed within the PERCEIVE project to predict and visualise colour changes in photosensitive materials exposed to light [17]. Its primary application lies in the preventive conservation of paintings and works on paper (PERCEIVE Scenario 2), although its framework can be extended to other photosensitive media such as textiles and photographs (Fig. 8).

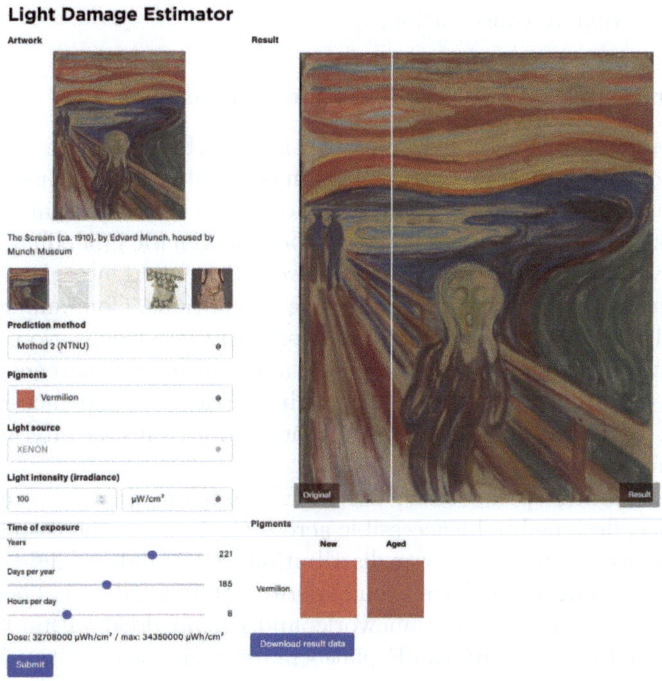

Fig. 8. Main interface of the Light Damage Estimator (LDE) tool, allowing users to explore and visualise the predicted effects of light exposure on objects over time.

The LDE integrates experimental data that allows the simulation of cumulative colour changes over various time horizons and lighting conditions. The tool accepts digital image inputs and provides visual outputs designed to support conservators, heritage scientists, and museum professionals in assessing lighting-induced risks and defining long-term preservation strategies.

In concept and purpose, the LDE builds upon and extends previous efforts such as the Canadian Conservation Institute's (CCI) Online Light Damage Calculator [2], which has been adopted within the preventive conservation community. The CCI tool was designed to estimate colour change resulting from light exposure based on photometric intensity (illuminance) measurements and exposure duration.

While the CCI calculator relies on illuminance and cumulative light dose to provide general light damage estimates, the LDE introduces a spectrally resolved and image-based predictive framework. By incorporating measured spectral power distributions and experimentally determined degradation kinetics of photosensitive materials, the LDE allows users to explore the impact of different lighting solutions and scenarios on the colour of an artwork, based on the predictable change in material-specific areas of the object. This evolution represents a significant advancement in predictive capability, bridging experimental data and real-world lighting conditions for more precise, actionable decision-making in preventive conservation.

6.2 User Requirements and Technical Approach

The development of the Light Damage Estimator (LDE) was guided by the specific needs of diverse user groups engaged in cultural heritage. Museum conservators and collection managers required an accessible means of quantifying object-specific risks from light exposure, moving beyond generic guidelines. Researchers and conservation scientists sought a framework that could integrate experimental data on pigment sensitivity into predictive simulations. Communication professionals and curators expressed the need for intuitive visualisations to facilitate dialogue with stakeholders and the public and ensure transparency in conservation decisions. Finally, the wider cultural industries and audiences benefit from a tool that translates complex scientific predictions into an interactive, understandable format.

From a technical perspective, the LDE is implemented as a modular web-based service, designed to be accessible across platforms without specialised software. It integrates spectral power distribution (SPD) data of light sources with colour change models that have been experimentally validated. The tool's architecture relies on open-source frameworks and is embedded within the microservice infrastructure of PERCEIVE platform, ensuring scalability and long-term maintainability.

The LDE emphasises interoperability with established standards. Colour predictions are expressed in the CIELAB 1976 colour space [1], allowing direct comparison with conservation guidelines and scientific literature. Image data management supports integration into museum imaging pipelines. In addition, the system is designed to interface with digital collection management systems and external repositories, thereby facilitating adherence to FAIR data principles(Findable, Accessible, Interoperable, Reusable).

6.3 Key Features and Functionalities

The LDE offers a set of core capabilities designed to support preventive conservation workflows. The central module enables the dynamic visualisation of predicted colour changes in artworks under user-defined lighting scenarios. The user can define a lighting scenario by selecting a light source, specifying the light intensity in illuminance [lx] or in irradiance [$\mu W/cm^2$] values, and defining the exposure time. For an exhibition, for example, the exposure time depends on the number of years, days per year, and hours per day defined by the user, as well as the total exposure dose (in either lx hours or $\mu Wh/cm^2$) (Fig. 9).

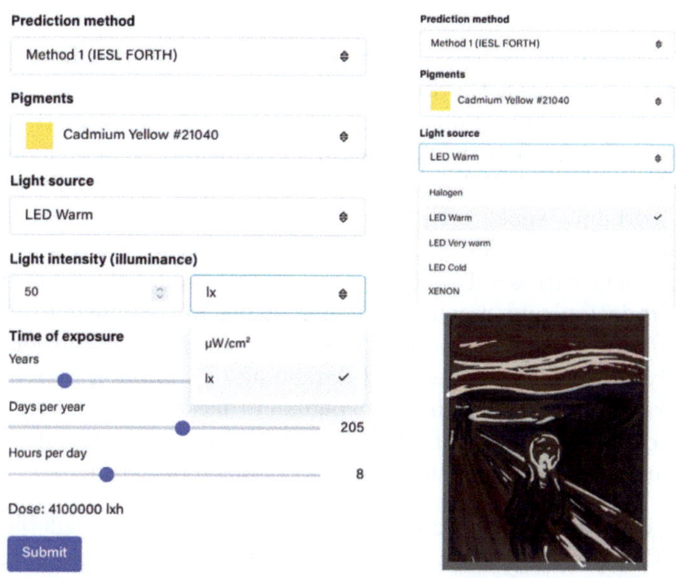

Fig. 9. Interface for selecting pigments and exposure parameters (left), along with light sources (top right), to simulate and visualise colour changes under varying lighting conditions. Pigment masks extracted from the artwork (bottom right) provide the spatial information for computing the colour changes.

A simulation engine combines pigment-specific sensitivity data with the spectral distribution of the selected light source to estimate the expected colour differences caused by light exposure in the areas of the artwork where the specific pigments are mapped. It then generates visual predictions, which are displayed directly on digital images of the artwork.

The current release of the tool offers two different simulation methods:

1. The first method integrated into the tool provides a mechanistic explanation of pigment colour change caused by light-induced chemical reactions (see Chap. 1, Sect. 3). It links light absorption to pigment degradation and correlates shifts in the CIELAB colour space with photochemical processes. The model incorporates the spectral power distribution of the light source, accounting for how different wavelengths influence degradation rates according to each pigment's absorbance profile. Long-term colour changes are expressed as a function of radiant exposure, enabling predictions to be

made about fading or hue shifts under specific lighting conditions. However, it assumes that light is the primary driver of pigment degradation and does not include secondary environmental factors such as temperature or humidity. Derived coefficients describe the quantitative relationship between colour change and radiant exposure, and are used to simulate future colour alterations. This photochemical method provides a predictive, physicochemical framework for assessing and managing light-induced deterioration in artworks.

2. The second method is based on a data-driven pigment photo-degradation model that operates in the hyperspectral domain to simulate light-induced colour changes in artworks (see Chap. 1, Sect. 3). This method combines hyperspectral imaging of the painting with artificially aged pigment mock-ups that replicate the original material palette. Using spectral linear unmixing in the absorbance space (–log R), the model identifies pigment abundance maps based on known spectral signatures and reconstructs the artwork by re-mixing them with aged spectra. This allows simulation of progressive pigment degradation under controlled light exposure. Applied to The Scream (ca. 1910), the model reproduced realistic pigment distributions consistent with analytical data and predicted darkening of vermilion regions with continued irradiation. Although some pigments lacked aging data, the model remains scalable—capable of integrating additional experimental datasets to refine future simulations of light-induced damage in paintings.

In terms of licensing and accessibility, the tool is made available as part of the PERCEIVE platform via a web interface. Currently, access is restricted to registered users within the consortium and associated institutions. However, its modular design anticipates eventual release under an open-access or community license to support open access and broader adoption. An authentication service ensures the confidentiality and secure handling of institutional data, as well as integration with project repositories.

The LDE supports a variety of standard data types and formats, including digital images of artworks (standard PNG formats), spectral power distribution (SPD) data of light sources and pigment artificial ageing datasets in a CSV format. The outputs of each user session consist of interactive visualisations and downloadable analytical reports (CSV), which include the necessary metadata for the visualisations produced interactively. This ensures compatibility with existing documentation practices. The tool is designed for integration with other platforms, most notably the PERCEIVE Knowledge Repository [15] and associated simulation services.

6.4 Use Scenarios and User Profiles

The LDE has been designed to support diverse conservation and curatorial contexts by offering flexible workflows. A typical use case begins with the selection of a light-sensitive object for which a display plan or storage strategy is required. A preliminary study using an analytical imaging technique, such as macroXRF mapping or HSI imaging, is required to acquire pigment maps. These pigment maps are then aligned with the uploaded reference colour image of the object within the LDE application.

The user then selects a light source from the library of available predefined or known spectral power distributions. The user then selects from the list of identified pigments on the object those for which a light damage estimation is necessary due to their high sensitivity. The tool allows for the estimation of single and multiple pigments. A growing library of data produced by accelerated ageing experiments on mock-ups supports the tool. The tool simulates cumulative colour changes under user-defined exposure conditions and visualises the results interactively. It enables comparative evaluations based on visualisations corresponding to different lighting scenarios. These outputs and predicted images can be archived for conservation records or shared in exhibition planning discussions.

The primary user profiles are:

- Museum professionals, such as lighting designers and conservators, who require reliable predictions to inform lighting strategies for the display or storage of individual artworks or collections.
- Researchers in conservation science, who use the tool to test and validate experimental models of pigment ageing.
- Communication and education specialists, who benefit from the intuitive visualisations to explain conservation challenges to broader audiences.
- The interested public who can better understand the effects of light on cultural heritage objects through guided demonstrations.

The LDE has been implemented in PERCEIVE Scenarios 2 and 3 ("Change of Colours in Paintings and Works on Paper" and "Fading Colours on Textiles") using several artworks as pilot applications. For example, simulations were performed on the sensitive pigments of specific artworks that formed PERCEIVE case studies. In the course of the project, the accelerated aging experiments provided the empirical basis for validating the models developed. The Light Damage Estimator (LDE) tool has been demonstrated and evaluated by both internal and external users to validate its performance and usability. A major demonstration campaign took place during the Digital Heritage 2025 Conference held in Siena, Italy (8–13 September 2025), where participants from research institutions, museums, and conservation laboratories tested the tool's functionality and provided feedback.

The application contexts of the LDE extend beyond museums to archives, libraries, and heritage sites where light-sensitive materials are exhibited. In exhibition lighting design, the tool supports evidence-based decision-making by balancing accessibility with preservation. In research contexts, it provides a framework for testing hypotheses about pigment stability under different lighting regimes. In educational contexts, it provides an interactive and visual means of raising awareness of preventive conservation.

6.5 Impact and Added Value

The LDE represents a significant advance in the field of preventive conservation by transforming laboratory-derived knowledge into a practical, accessible tool for everyday decision-making [22]. By embedding experimentally validated pigment sensitivity data into a user-friendly interface, the tool has bridged the gap between fundamental research and applied conservation practice. This enables museums and research institutions to move beyond generic light-exposure guidelines and adopt evidence-based, object-specific strategies for display and preservation.

The benefits of LDE are wide-reaching. In terms of knowledge creation, for example, it enables systematic documentation of predictable light-induced colour changes, generating new datasets that enrich both scientific research and conservation records alike. It also fosters collaboration between scientists, conservators, and curators, as the tool's visual simulations provide a basis for interdisciplinary dialogue. The tool offers researchers a structured environment in which to validate and extend predictive models, and provides curators with a transparent means of communicating conservation risks to decision-makers and the public.

The LDE also improves accessibility and decision-making processes. Its visualisations enable non-specialists to grasp the potential consequences of light exposure, supporting transparent exhibition planning and risk communication.

In terms of community building, the LDE is already fostering a user network within the PERCEIVE consortium, which includes leading museums and research institutions. Its modular web-based design will facilitate expansion to a broader community of practice in the future, enabling the sharing of workflows, case studies, and training initiatives.

Preliminary uptake indicators point to strong interest from PERCEIVE Scenario 2 institutions, where the tool has been piloted on paintings and works on paper. Although it is still under evaluation, the early adoption of the LDE suggests its potential for integration into routine conservation workflows. In the long term, impact is expected to be reflected in measurable outcomes, such as the prevention of sensitive pigment alteration through informed lighting scenarios, the promotion of object-specific lighting strategies, and the incorporation of the LDE into professional training for conservators and museum staff.

6.6 Sustainability and Future Developments

The sustainability of the Light Damage Estimator (LDE) is ensured through its integration within the PERCEIVE platform, which provides institutional hosting and ongoing technical support by a consortium of leading research organisations. Designed as a modular and extensible service, the tool can evolve in response to user feedback and advances in conservation science.

Planned updates and new features include the systematic expansion of the pigment sensitivity database through additional accelerated ageing studies, the incorporation of artificially aged pigment data, and the enhancement of visualisation options such as uncertainty mapping and side-by-side scenario comparisons.

Future developments will also focus on the import of custom spectral profiles for light sources, the automated reporting of cumulative damage, and the export of simulation results for conservation documentation and analysis purposes.

The sustainability model is based on institutional hosting during and beyond the PERCEIVE project, with provisions for a community-driven development model as the tool matures. This will be supported by training workshops, user guides, and open-access documentation, encouraging widespread adoption. A growing user base will further strengthen sustainability by contributing case studies, validation data, and iterative feedback.

In terms of scalability and long-term maintenance, the LDE benefits from its API-based, web-service architecture, which facilitates interoperability with external repositories, digital collection management systems, and other heritage science tools. Compliance with widely adopted standards, including CIE colour metrics for predictive modelling and IIIF protocols for image interoperability, ensures compatibility with emerging infrastructures in digital heritage.

Finally, the LDE is inherently customisable and extensible. Its modelling framework can accommodate new pigment sensitivity datasets, integrate multi-factorial ageing parameters, and expand to additional visualisation modes. This flexibility guarantees that the tool will remain scientifically robust and practically relevant, positioning the LDE as a long-term reference resource for managing light-sensitive collections well beyond the lifetime of the PERCEIVE project.

7 Deep Degreening Tool

7.1 General Description and Scope

Autochromes represent one of the pioneering advancements in colour photography and hold a vital place in the history of visual arts. Their unique composition makes them exceptionally fragile and sensitive to light, resulting in irreversible damage when they deteriorate. In our project, PERCEIVE [16], we developed AI models that are able to restore digitised autochromes that display greening defects. The de-greening tool uses the this AI service to de-green the input defected image (see Fig. 10).

7.2 User Requirements and Technical Approach

The objective is to restore digitised autochrome photographs that exhibit greening defects completely or partially. In case of partial removal curators and designers should be able to address the remaining image issues after the green regions are removed in a shorter time-frame.

To achieve this, we generate synthetic greening defects [20] through image processing, as we lack reference or ground-truth images for restoration. Our analysis of genuine defects shows they typically manifest as small, point-like stains or larger leakage-style areas, both featuring multi-ringed green halos with subtle orange borders. We simulate these effects by randomly selecting whether an image receives only spot defects, only large area defects, or a combination

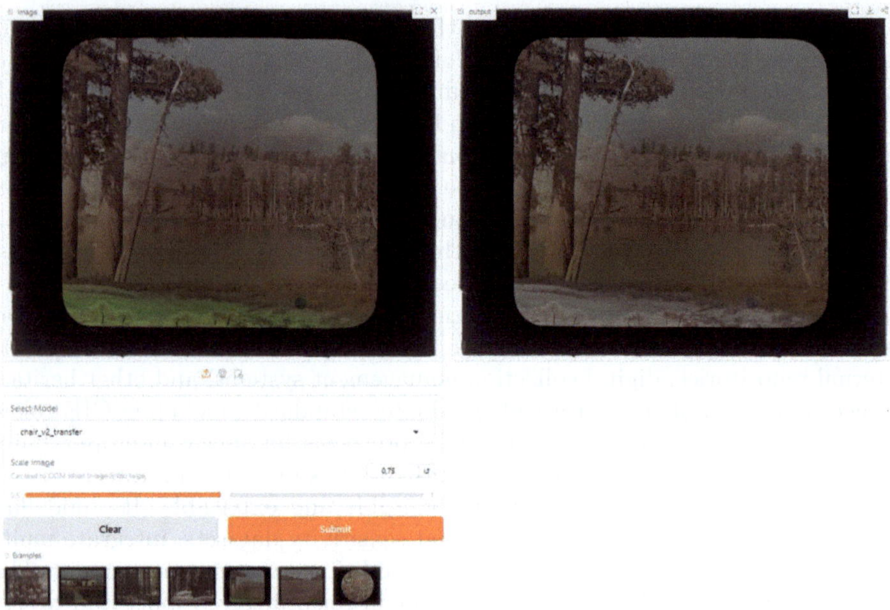

Fig. 10. User interface of the deep de-greening tool.

of both. This process yields a diverse dataset of images with realistic greening defects, which serve as a foundation for training our restoration networks.

7.3 Key Features and Functionalities

Our de-greening tool offers several model checkpoints, each trained on different loss functions and slightly varied datasets, allowing users to choose the one that best suits their needs. Users can easily select their preferred model checkpoint and specify the desired scale for the output image (see Fig. 10). Once these options are set, they can run the algorithm to generate a de-greened image. This flexibility ensures that users can tailor the restoration process to achieve optimal results based on their specific requirements. The interface is designed to be user-friendly, making it simple for anyone to restore their autochrome photographs effectively.

7.4 User Scenarios and User Profiles

One scenario involves a museum curator who discovers an old autochrome photograph with greening defects. To restore the image, the curator accesses our restoration tool, selects a suitable model checkpoint based on prior successes, and specifies the desired output scale before running the algorithm to de-green the image. After reviewing the results, if satisfied, the curator may perform additional post-processing with image editing tools or send the image to a designer or specialist curator for further refinement.

Another scenario features an amateur photographer who encounters a digitised autochrome in their collection that has developed greening defects. They use the same restoration tool to select a model checkpoint and set the output scale before running the algorithm. Upon evaluating the de-greened image, they may enhance it further with their favorite editing software or seek feedback from a professional.

In both cases, the tool can display the original image alongside the restored version whenever shared, helping to preserve the integrity of the historical photographs and ensuring sensitivity to their cultural heritage significance. These scenarios highlight the needs of both museum curators, who prioritise preservation and integrity, and hobbyists, who seek user-friendly tools that allow for experimentation while understanding the importance of historical context.

7.5 Impact and Added Value

The restoration tool helps in the preservation of autochrome photographs, offering both museum curators and amateur photographers a means to effectively address greening defects while maintaining the integrity of these historical artifacts. For hobbyists, the tool provides a user-friendly platform to experiment with image restoration, fostering a deeper appreciation for historical photography. By allowing users to view the original and restored images side by side, the tool promotes awareness of the importance of preserving the authenticity of historical works. Overall, the tool adds substantial value by streamlining the restoration process, making it accessible to a broader audience, and promoting sensitivity to the historical context of autochromes.

7.6 Sustainability and Future Developments

The sustainability of our restoration tool lies in its capability to adapt and evolve alongside advancements in technology in the field of image processing, AI and also user needs. Future developments can focus on addressing additional autochrome defects, such as orangings, utilising the same methodology as we did for greening. Moreover, improving the dataset is a critical aspect of future work. Involving a larger and more diverse set of images will help refine the training of our AI models, leading to more accurate and effective restoration outcomes.

8 Text2Autochromes

8.1 General Description and Scope

The previous section explains how we can remove greening defects from digitised autochromes by creating synthetic greening defects using image processing techniques. In this section, we utilise generative AI to generate autochrome images from text, including those with greening defects. This innovative method enables amateur users to create autochromes based on textual data while allowing AI experts to generate defects in autochromes for which no references exist. Our new approach employs generative models to create images that capture the unique qualities of autochromes. This also involves simulating common defects such as greening and cracking to develop machine learning methods for restoring real autochrome images with such defects. By generating synthetic defects, such as images with greening issues, we can create a dataset of defected and non-defected autochrome image pairs. This will allow us to train an AI model that learns to effectively remove these greening defects, potentially restoring the integrity of autochrome images.

8.2 User Requirements and Technical Approach

There is no reference available for the greening defects present in digitised autochrome images, making the synthetic simulation of these defects crucial. Text2Autochromes [12] leverages generative AI to create such defects from textual descriptions, fulfilling the need for data that AI experts require. This tool also allows designers and amateur users to create content in the style of autochrome images. However, it is important that such content is explicitly labeled as AI-generated to address ethical concerns regarding AI generated images.

To train a diffusion model to generate distinctive autochrome style, we collected 420 high-resolution scans from glass-plate photographs [23], each paired with its original caption. These images were resized for GPU memory and the prompts were slightly rephrased to enhance the training dataset. We then fine-tuned the FLUX.1-dev model [13] using Low-Rank Adaptation (LoRA) [11]. Throughout the training process, we closely monitored sample outputs and computed CLIPScore, a text–image compatibility metric, using varied prompts to detect any signs of overfitting.

8.3 Key Features and Functionalities

The tool provides utilises the AI backend service for generating autochromes from user prompts. Users can create images in various aspect ratios to suit their specific needs. Additionally, the system allows for the selection of different checkpoints, enabling users to tailor the generation process according to their requirements (see Fig. 11). Key parameters such as guidance scale and the number of inference steps can be adjusted to influence the output quality and style. Furthermore, users can utilise a seed value to ensure consistent generation across multiple iterations with the same text prompt, enhancing the coherence and reliability of the generated autochrome images.

8.4 User Scenarios and User Profiles

This tool is designed to cater to a diverse range of users, each with unique needs and objectives. Amateur photographers are interested in experimenting with

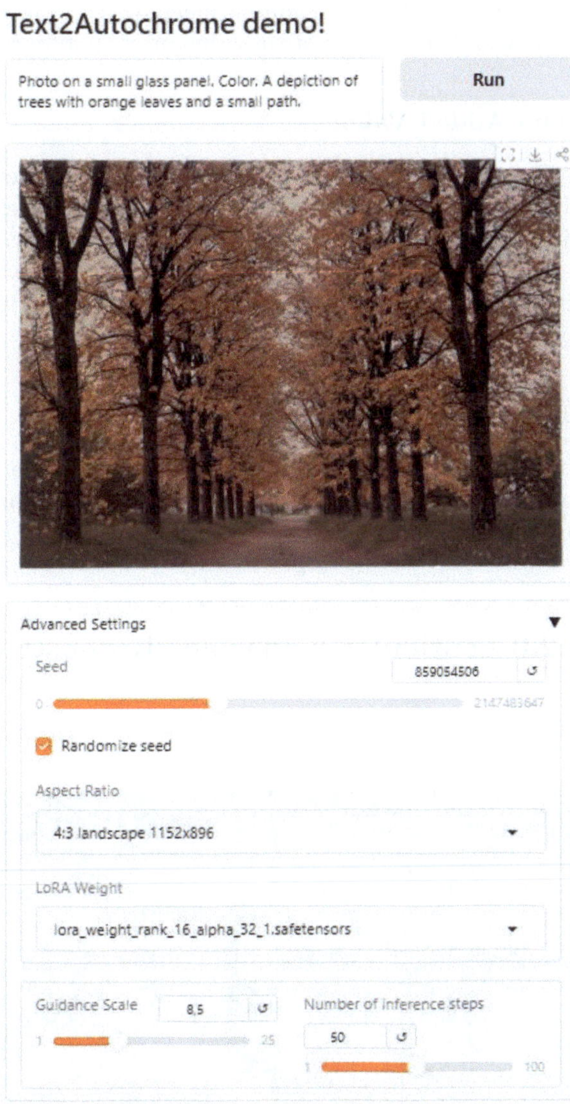

Fig. 11. User interface of the Text2Autochrome tool.

autochrome styles to enhance their photography; they can input text prompts describing their desired scenes and receive AI-generated images that mimic the distinctive qualities of autochromes. Graphic designers looking to incorporate vintage style aesthetics into their projects can use the tool to generate autochrome-style images, adjusting parameters like aspect ratio and guidance scale to achieve the desired visual effect. Additionally, AI experts can utilise this tool for generating synthetic defects in autochrome images, such as greening or cracking, which are essential for training machine learning models aimed at restoring and enhancing the quality of these historical photographs.

8.5 Impact and Added Value

The generative AI tool specifically enables the creation of synthetic defects where no ground truth exists for the restored image, addressing the lack of reference for greening and other imperfections in autochrome photography. This capability allows AI experts to generate training data essential for developing machine learning models focused on image restoration, significantly advancing the field of digital preservation.

8.6 Future Developments

Currently, the tool works well for generating autochrome-style images, particularly in landscapes. However, further training with better datasets is needed to effectively address a wider range of subjects, including human figures and other diverse scenes.

9 DoLCE 2.0 Lenticular Kodacolor Reconstruction

(See Fig. 12).

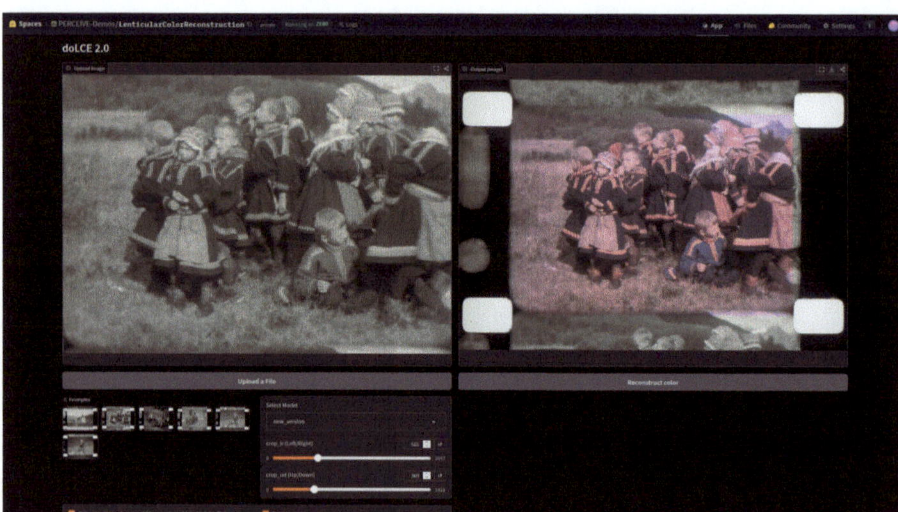

Fig. 12. User interface of the Lenticular Colour Reconstruction Tool.

9.1 General Description and Scope

The Lenticular Colour Reconstruction Tool (doLCE 2.0) is a software framework developed within the PERCEIVE project to recover the original colour information embedded in early lenticular films such as Kodacolor. It extends the pioneering doLCE method originally conceived at the University of Basel, integrating recent advances in deep learning and user-guided refinement to achieve more faithful digital restorations of these historically significant materials. The tool enables cultural heritage professionals, researchers, and audiovisual archivists to reconstruct full-colour digital images from monochrome lenticular film scans, thereby preserving rare early colour cinematography and ensuring its accessibility for future generations.

doLCE 2.0 forms part of PERCEIVE Scenario 4 (Colour Reconstruction in Historical Film), where it serves both as a research instrument and a practical preservation solution. By combining computational imaging, neural modelling, and heritage expertise, the system bridges the gap between historical technology and contemporary digital conservation practice.

9.2 User Requirements and Technical Approach

The development of doLCE 2.0 has been guided by the needs of film archives, conservation scientists, and audiovisual restoration specialists seeking robust and accessible means to decode lenticular colour information without dependence on obsolete optical projection equipment. Archival institutions require a reproducible and transparent digital workflow that transforms high-resolution film scans into perceptually accurate colour reconstructions. Researchers demand methodological rigour and the ability to compare algorithmic outputs with historical documentation and physical evidence. Curators and educators, in turn, benefit from visual reconstructions that communicate the aesthetic qualities of early colour film to contemporary audiences.

From a technical perspective, doLCE 2.0 implements a modular, data-driven pipeline that combines traditional signal processing with deep neural network architectures. The tool detects the microscopic lenticular pattern embossed on the film base and translates its spatially encoded silver densities into RGB colour channels. The updated version introduces adaptive finetuning mechanisms based on user-validated training data, substantially improving performance on heterogeneous or degraded source material. Implementation within the PERCEIVE microservice infrastructure ensures interoperability, scalability, and long-term maintainability. Data handling adheres to FAIR principles, supporting integration with film digitisation workflows and institutional repositories.

9.3 Key Features and Functionalities

doLCE 2.0 provides a suite of analytical and visual modules for the decoding and colour reconstruction of lenticular film material. Key features include:

- Automated lenticule detection and alignment, using convolutional neural networks finetuned on representative training datasets.
- User-guided finetuning workflow, enabling experts to iteratively improve model accuracy through manual correction or automatic selection of plausible frames.
- Colour reconstruction and calibration, transforming the encoded silver densities into RGB images consistent with known spectral characteristics of Kodacolor and related processes.
- Export functionality, generating downloadable, colour-accurate frames ready for archival, analysis, or visualisation.

Access to the tool is provided through the PERCEIVE web interface for registered users, with planned future release under an open or community license. Its modular design ensures compatibility with complementary services within the platform, including the PERCEIVE Knowledge Repository and colour analysis tools.

9.4 Use Scenarios and User Profiles

The tool supports a range of professional and research contexts in film preservation, computational imaging, and historical reconstruction. Typical workflows include:

- Film archivists and conservators, who apply doLCE 2.0 to digitally restore lenticular films such as Kodacolor, ensuring faithful colour recovery without analogue projection.
- Researchers in heritage science and image processing, who use the system to test algorithmic methods for colour decoding and to study historical optical encoding techniques.
- Curators and educators, who employ reconstructed footage to illustrate the evolution of early colour cinema and its technological heritage.
- Public audiences, who experience re-created colour sequences that were previously inaccessible or misidentified as black-and-white materials.

A representative case study within PERCEIVE involved the reconstruction of a 1930 travelogue to the Norwegian fjords by amateur filmmaker Arnold Hennings. This 16 mm Kodacolor segment presented severe decoding challenges due to degraded lenticular patterns. By finetuning the deep-learning model with a dataset of manually corrected and visually validated frames, doLCE 2.0 achieved stable and historically plausible colour reproduction. The iterative workflow—combining expert correction and automatic frame selection—demonstrated significant gains in accuracy, producing a coherent and visually credible restoration of the original colour imagery.

9.5 Impact and Added Value

doLCE 2.0 represents a major step forward in the digital preservation of early colour film heritage. By transforming high-resolution scans into authentic colour reconstructions, the tool bridges the technological gap between obsolete lenticular systems and modern digital imaging. It contributes to the rediscovery and scholarly understanding of the origins of colour cinema, enriching both film history and conservation practice.

For the heritage community, doLCE 2.0 offers several tangible benefits:

- Protect vulnerable acetate film materials by replacing mechanical projection with a fully digital decoding process.
- Generate standardised digital outputs that can be archived, studied, and shared between institutions.
- Promotes interdisciplinary collaboration among computer scientists, conservators, and historians, fostering a shared methodological framework for colour reconstruction.
- Enhance public engagement by making early colour footage visible again, providing a unique window into historical visual culture.

By embedding advanced machine-learning techniques in an accessible and transparent interface, doLCE 2.0 exemplifies how data-driven approaches can be responsibly applied to cultural heritage. It transforms technical research into a practical tool for the safeguarding and dissemination of early colour cinema.

9.6 Sustainability and Future Developments

The sustainability of doLCE 2.0 is ensured through its integration within the PERCEIVE digital infrastructure, supported by institutional hosting and collaborative maintenance among project partners. The tool's modular, open-source architecture enables continuous refinement through community contributions and ongoing research in computational imaging.

Planned developments include expanding the training dataset to encompass additional lenticular film types (such as Agfacolor and Keller-Dorian), integrating metadata to support provenance tracking and uncertainty assessment, and introducing visual analytics tools for comparing reconstruction hypotheses. Future updates will also focus on improving user accessibility through cloud-based processing for large film sequences and export options aligned with audiovisual preservation standards (e.g., DPX, FFV1). Additional features will include the generation of downloadable colour frames and reconstruction logs for documentation and scholarly analysis, as well as batch processing and interactive visual previews that enable users to apply reconstruction settings across complete film sequences and review results dynamically. Through its combination of scientific rigour, digital preservation value, and community-driven design, doLCE 2.0 is positioned to become a long-term reference tool for the recovery and study of lenticular colour films, extending the legacy of PERCEIVE's commitment to the sustainable preservation of our visual heritage.

Acknowledgments. This study was funded by the European Union (grant number 101061157, Project PERCEIVE).

Author contributions. Sophia Sotiropoulou contributed to Sect. 1; Marios Pitikakis and Sophia Sotiropoulou contributed to Sect. 2; Paschalis Penteleris and George Papadopoulos contributed to Sect. 3; Bruno Fanini, Marcello Massidda and Petros Ioannis Stavroulakis contributed to Sect. 4; Saptarshi Neil Sinha contributed to Sect. 5; Panagiotis Siozos, Irina-Mihaela Ciortan, Petros Ioannis Stavroulakis and Sophia Sotiropoulou contributed to Sect. 6; Saptarshi Neil Sinha and Paul Julius Kühn contributed to Sect. 7; Saptarshi Neil Sinha and Giorgio Trumpy contributed to Sect. 8; Giorgio Trumpy and Saptarshi Neil Sinha contributed to Sect. 9.

Disclosure of Interests. The authors have no competing interests to declare that are relevant to the content of this article.

References

1. Beleuchtungskommission, I. (ed.): Colorimetry. No. 15 in Publication/CIE, Comm. Internat. de l'éclairage, Wien, 3rd edn. (2004)
2. Canadian Conservation Institute: Online light damage calculator (2025). https://app.pch.gc.ca/application/cdl-ldc/description-about.app?lang=en. Accessed 23 Oct 2025
3. European Commission: Turning FAIR into Reality: Final Report and Action Plan from the European Commission Expert Group on FAIR Data. Publications Office of the European Union (2018). https://doi.org/10.2777/1524
4. Fanini, B., Ferdani, D., Demetrescu, E.: Temporal lensing: an interactive and scalable technique for Web3D/WebXR applications in cultural heritage. Heritage **4**(2), 710–724 (2021)
5. Fanini, B., Ferdani, D., Demetrescu, E., Berto, S., d'Annibale, E.: Aton: An open-source framework for creating immersive, collaborative and liquid web-apps for cultural heritage. Appl. Sci. **11**(22), 11062 (2021)
6. Fanini, B., Gosti, G., Amadei, C.F., Menconero, S.: Shaping a Web3D framework for different scientific communities: ATON as a service in H2IOSC. In: Campana, S., et al. (eds.) Digital Heritage. The Eurographics Association (2025). https://doi.org/10.2312/dh.20253055
7. Fanini, B., et al.: Interactive and immersive discovery of diagnostic processes on multi-layered 3D collections on the web: the MuLaX tool. In: Campana, S., et al. (eds.) Digital Heritage. The Eurographics Association (2025). https://doi.org/10.2312/dh.20253102
8. Ferdani, D., et al.: Brancacci Chapel in Florence: surveying and real-time 3D simulation for conservation and communication purposes. Int. Arch. Photogramm. Remote. Sens. Spat. Inf. Sci. **48**, 535–540 (2023)
9. Ferdani, D., Barandoni, C., Bonifazi, F., Iannaccone, R., Magrini, D.: Multiband photogrammetry for multispectral 3D reconstruction of statuary. J. Cult. Herit. **76**, 74–85 (2025)
10. Hahnel, M., Treadway, J., Fane, B., et al.: The State of Open Data Report 2018. Digital Science (2018). https://doi.org/10.6084/m9.figshare.7195058.v2
11. Hu, E.J., et al.: LoRA: low-rank adaptation of large language models (2021). https://arxiv.org/abs/2106.09685
12. Kühn, P.J., Sinha, S.N., Nguyen, D.A., Horst, R., Kuijper, A., Fellner, D.W.: Text2Autochrome: text guided autochrome synthesis using generative models. In: Campana, S., et al. (eds.) Digital Heritage. The Eurographics Association (2025). https://doi.org/10.2312/dh.20253061

13. Labs, B.F., et al.: FLUX.1 Kontext: flow matching for in-context image generation and editing in latent space (2025). https://arxiv.org/abs/2506.15742
14. Padfield, J., Sotiropoulou, S., Fremout, W., Schmidle, W.: E-RIHS Heritage Science Interoperability (v0.5-beta) (2024). Zenodo: https://doi.org/10.5281/zenodo.10522529 GitHub: https://github.com/E-RIHS/hs-interoperability/
15. PERCEIVE Consortium: Perceive knowledge repository (2025). https://perceive-data.iesl.forth.gr/
16. Pescarin, S., et al.: Perceptive enhanced realities of coloured collections through AI and virtual experiences. In: Campana, S., et al. (eds.) Digital Heritage. The Eurographics Association (2025). https://doi.org/10.2312/dh.20253350
17. Pescarin, S., et al.: Perceptive enhanced realities of coloured collections through ai and virtual experiences. In: Digital Heritage (Proceedings of the Eurographics Workshop on Graphics and Cultural Heritage). Eurographics Association (2025). https://doi.org/10.2312/dh.20253350
18. Pietroni, E., Chirivì, A., Fanini, B., Bucciero, A.: An innovative approach to shape information architecture related to ancient manuscripts, through multi-layered virtual ecosystems. From Codex4D to dataspace project. In: De Paolis, L.T., Arpaia, P., Sacco, M. (eds.) International Conference on Extended Reality. pp. 247–267. Springer, Cham (2023). https://doi.org/10.1007/978-3-031-43404-4_16
19. Robinet, F., et al.: glTF: designing an open-standard runtime asset format. In: GPU Pro 360 Guide to 3D Engine Design, pp. 243–260. AK Peters/CRC Press (2018)
20. Sinha, S.N., Kuehn, P.J., Koppe, J., Kuijper, A., Weinmann, M.: Neural restoration of greening defects in historical autochrome photographs based on purely synthetic data (2025). https://arxiv.org/abs/2505.22291
21. Sinha, S.N., Kühn, P.J., Rojtberg, P., Graf, H., Kuijper, A., Weinmann, M.: Semantic stylization and shading via segmentation atlas utilizing deep learning approaches. In: Smart Tools and Applications in Graphics - Eurographics Italian Chapter Conference. The Eurographics Association (2024https://doi.org/10.2312/stag.20241352
22. Siozos, P., et al.: Light source classification and colour change modelling for understanding and predicting pigments discolouration. J. Phys, Photonics (2025)
23. Taylor, H.A.: Harold Taylor slide collection. California Revealed (1896). https://californiarevealed.org/search?search_api_fulltext=&f%5B0%5D=search_page_series_title:Harold%20Taylor%20Slide%20Collection. Accessed 26 Nov 2024
24. Tominski, C., Gladisch, S., Kister, U., Dachselt, R., Schumann, H.: Interactive lenses for visualization: an extended survey. Comput. Graph. Forum **36**, 173–200 (2017)
25. Wilkinson, M.D., Dumontier, M., Aalbersberg, I.J., et al.: The FAIR Guiding Principles for scientific data management and stewardship. Sci. Data **3**, 160018 (2016). https://doi.org/10.1038/sdata.2016.18
26. Zenodo CERN (2025). https://zenodo.org/ and https://about.zenodo.org/

Design Resources to Enhance Access to Coloured Collections Through Interactive Experiences

Marcello Massidda[1,2], Birgitte Aga[3], Cristiana Barandoni[1], Federica Bonifazi[1,2], Ivana Cerato[1], Arthur Clay[4], Enzo D'Annibale[1], Bruno Fanini[1], Daniele Ferdani[1], Marialucia Giacco[5], Roberta Iannaccone[1], Donata Magrini[1], Caterina Serena Martucci[5], Nynne Ryhl Olsson[3], Irina Crina Anca Sandu[3], Samuele Spotti[1], Laura Travaglini[1], Giorgio Trumpy[6], Sofia Pescarin[1], and Manuele Veggi[1,2](✉)

[1] Institute of Heritage Science, National Research Council, Florence & Rome, Italy
{marcello.massidda,cristianabarandoni,federicabonifazi,ivana.cerato, enzo.dannibale,bruno.fanini,daniele.ferdani,robertaiannaccone,donata.magrini, samuelespotti,lauratravaglini,sofia.pescarin,manueleveggi}@cnr.it, {marcello.massidda,federica.bonifazi,manuele.veggi}@uniroma1.it
[2] Department of Antique Studies, Sapienza University of Rome, Rome, Italy
[3] MUNCH Museum, Oslo, Norway
{birgitte.aga,nynne.olsson,irina.sandu}@munchmuseet.no
[4] Hochschule Luzern (HSLU), Luzern, Switzerland
arthur.clay@hslu.ch
[5] National Archaeological Museum (MANN), Naples, Italy
{marialucia.giacco,caterinaserena.martucci}@cultura.gov.it
[6] Norwegian University of Science and Technology, Gjøvik , Norway
giorgio.trumpy@ntnu.no

Abstract. This chapter presents the design strategies and experimental outcomes of the PERCEIVE project, aimed at enhancing access to coloured heritage through interactive and hybrid experiences. Grounded in the principles of authenticity, sense of care, and participation, the research translates studies on colour perception, digital reconstruction, and conservation into reusable design resources for museums. Four prototypes—*The Gifts of Isis*, *Echoes of Loss*, *Tiny Conservators*, and the *Open Space Museum*—exemplify these principles in practice. Central to this process is also the PERCEIVE Design Toolbox, an integrated framework supporting co-design, prototyping, exhibition planning and evaluation. Lastly, the demonstrators test further methods to foster participatory, embodied, and sustainable approaches to coloured collections.

Keywords: Concept-oriented Design · Collaborative Design · Participation · Coloured Collections

I. M. Ciortan et al. (eds.), Chromatic Visions, LNCS 16450, pp. 97–146, 2026.
https://doi.org/10.1007/978-3-032-17700-1_3

1 Introduction

This chapter presents the design strategies and experimental outcomes developed within the PERCEIVE project to enhance access to coloured collections through interactive and hybrid experiences. The work translates research on colour perception, digital reconstruction, and conservation into a series of design resources - prototypes, demonstrators, and structured templates - conceived for reuse across different museum contexts and audiences. A shared prototype template documents each project's rationale, learning goals, interaction models, technologies, and reusability criteria, ensuring that these experiences can be replicated and adapted beyond the scope of the project. At the core of this process lies a design philosophy articulated around three interdependent concepts: authenticity, sense of care, and participation.

- The **Authenticity** principle guides the creation of immersive and collaborative environments that evoke trust, reflection, and awareness in digital encounters with heritage, making visible the scientific processes and uncertainties behind reconstruction.
- The **Sense of Care** principle encourages visitors to develop empathy toward the fragility of colours and materials and to perceive conservation as a shared, ethical, and aesthetic responsibility.
- The **Participation** principle expands the museum beyond its physical boundaries, promoting co-creation, inclusion, and the active involvement of diverse publics in the encounter and interpretation of Cultural Heritage.

Within this conceptual framework, the chapter discusses four complementary outcomes developed as part of PERCEIVE's applied research. We wanted, in fact, to apply our theoretical research into real case studies. For this reason, we have identified suited application contexts, such as the archaeological context of the Iseum in Pompeii with its statues and frescoes conserved at National Archaeological Museum of Naples (MANN), and the Scream by Edward Munch conserved at the MUNCH museum, and developed specific experiences for the visitors, guided by the above-mentioned principles. ***The Gifts of Isis*** invites visitors to explore how colour once shaped the statues and walls of the Temple of Isis complex in Pompeii, offering a dynamic comparison between the present appearance of the artefacts and their possible ancient polychromy. ***Echoes of Loss*** translates scientific and artistic knowledge into embodied and emotional experiences, turning diagnostics and imaging into accessible, interactive installations that reveal the tools, steps, and methods of conservation. These first two prototypes are part of the exhibition *Perceive Isis' Colours*, launched at the Digital Heritage Expo in September 2025 [5]. Furthermore, ***Tiny Conservators*** engages families and schools through a cooperative 2D game that transforms conservation variables into intuitive, collaborative tasks, providing measurable outcomes for learning and participation. Finally, the ***Open Space Museum*** (OSM) relocates these ideas into the urban environment through modular, phygital installations that redefine the visitor as co-creator of meaning and extend the museum experience into public space.

Implementation across these case studies is supported by an integrated set of design tools, gathered in the **PERCEIVE Design Toolbox**. This includes (i) the **CoDesign Tool** for collaborative ideation; (ii) the **Prototyper** for rapid visualisation and simulation, and (iii) the **Experiment Planner** for testing and evaluating.

The following chapter sections are structured as follow: Sect. 1 introduces these design tools and methodologies; Sect. 2 presents the prototypes and exhibitions designed around the case-study logic; and Sect. 3 describes the demonstrators developed from these prototypes, highlighting how they contribute to new, participatory forms of chromatic heritage experience.

2 PERCEIVE Design Toolbox

The PERCEIVE Design Toolbox is a collection of easy-to-use, open-access resources developed to support curators, designers, educators, and students in the creation of cultural heritage experiences. It was designed in response to a widespread need in the field for more practical, intuitive tools that can also be used by non-specialist professionals, making it easier to collaborate across interdisciplinary teams and manage the increasing complexity of hybrid, value-driven experiences. Freely available through the PERCEIVE platform, the toolkit includes dedicated instruments for each key phase of the design process:

- the CoDesign Tool, a structured methodology for collaborative ideation using cards, templates, and scenario-building exercises;
- the Prototyper, a prototyping environment used between concept and first release to create 3D simulations of interaction flows and preliminary setups combining content and technology;
- the Experiment Planner, which supports the alignment of content, logistics, and audience engagement when transforming prototypes into public exhibitions and demonstrators.

Each of these tools will be described in detail in the dedicated subsections that follow.

2.1 CoDesign Tool: A Card-Based Framework for Interactive and Collaborative Design in Cultural Heritage

The CoDesign Tool is the first entry point in the PERCEIVE Design Toolbox. It is the tool specifically developed to support the initial phase of the creative process, when teams begin shaping their vision for a cultural heritage experience. This early ideation step is often one of the most difficult to manage, particularly within the cultural heritage domain, where interdisciplinary teams are the norm. Team members typically come from a variety of professional backgrounds and may not share the same familiarity with design methods or collaborative tools.

To address this, the CoDesign Tool offers a structured and easy-to-follow workflow, composed of simple, sequential steps that reflect a typical design process. Its format takes inspiration from card games, making it not only accessible but also engaging and playful. As such, no prior experience with design is required.

The tool can be used by cultural professionals, curators, designers, students, educators, and researchers alike. It is suitable for various contexts, including workshops, internal planning meetings, and co-creation sessions with institutional or external stakeholders.

Background. The CoDesign Tool builds upon the foundations of VisitorBox, a collaborative design framework originally developed within the EU-funded GIFT project [19]. While highly valuable, VisitorBox revealed several limitations when tested in real-world settings. These included a logistically demanding workflow and a lack of explicit focus on emotional and cognitive goals. Such shortcomings are especially relevant today, as emotional resonance and reflective engagement are increasingly recognised as essential to designing meaningful cultural heritage experiences that align with the expectations of contemporary audiences.

Rather than discard the tool, the PERCEIVE project recognised its potential and undertook a complete redesign [50]. This process preserved VisitorBox's core strengths while addressing its most pressing limitations. The development was highly iterative: each new version of the tool was tested through co-design workshops and refined based on direct feedback from cultural professionals, students, creative practitioners, and PERCEIVE project partners, who actively participated in and benefited from these collaborative sessions.

One of the most significant additions introduced during the redesign process was a new stage focused on cognitive-emotional goals (The Cognitive Goal Stage). This responded to the above-mentioned need for cultural experiences to connect with visitors not only intellectually but also emotionally. After an extensive review of key themes in the field, two values were selected for deeper exploration: Authenticity and Sense of Care. These concepts were also investigated through parallel theoretical work within the project, but the CoDesign Tool provided a space to make them actionable. Each value is represented by a dedicated deck of cards, developed as research-informed prompts that help teams reflect on the emotional and cognitive weight of their design choices early in the process.

Structure and Workflow. The CoDesign Tool is structured around two main phases, each supported by a dedicated collaborative board. These boards represent the two fundamental moments of the design process: problem-setting and problem-solving. The first board is dedicated to building the Design Brief, while the second focuses on generating and refining creative ideas.

Each phase consists of a sequence of modular stages, represented by individual canvases that are stacked vertically on the boards. Teams work through the canvases one at a time, and as each stage is finished, the canvas can be peeled away, gradually revealing a fully populated board underneath. This stacking layout offers a clear sense of progression while keeping the session compact, easy to navigate, and less physically demanding than traditional multi-board setups.

To support flexibility, each board also features a dedicated space, called Drawer of Alternatives, where teams can temporarily place discarded cards or ideas that might be revisited later.

Phase 1: Building the Design Brief

The first phase, and corresponding board, helps teams establish a shared foundation for their project. It guides them through four key stages:

- Context: Teams define the broader framework of their design challenge by selecting a target audience and identifying the experience's star assets.
- Institutional Goals: The discussion centres on what the hosting organisation aims to achieve, ensuring alignment with its broader mission and strategic objectives.
- Audience Goals: Drawing on dedicated card decks focused on Authenticity and Sense of Care, teams explore the emotional and cognitive values the experience should address.
- Design Brief: The final canvas brings together all prior elements, with teams defining key design parameters such as barriers, motivations, capabilities, and available devices (see Fig. 1b).

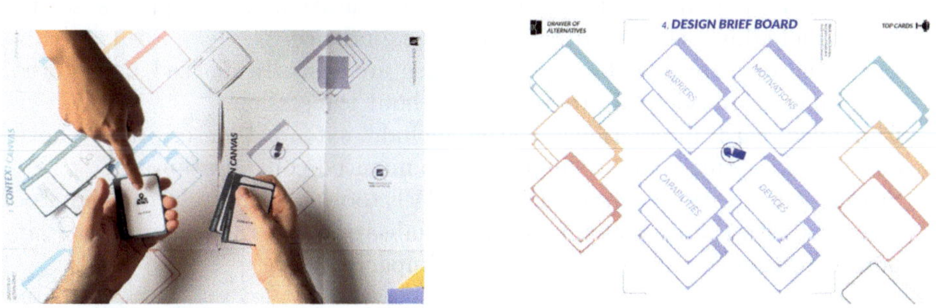

(a) (b)

Fig. 1. PERCEIVE CoDesign Tool enables collaborative design processes through a game-like approach, composed of cards (a) and paper boards (b)

Phase 2: Ideation and Storyboarding The second phase supports the more creative and exploratory part of the process. It includes the following, final stages:

- Ideation: Using thematic prompt cards, teams generate multiple possible design directions. This stage can be repeated as needed until a promising concept is selected.

- Disruption (reintroduced as part of the redesign): The selected idea is tested against scenario-based cards that challenge its resilience, ethical soundness, and feasibility. Depending on the outcome, teams may make minor refinements or revisit earlier decisions.
- Storyboarding: The refined concept is developed into a simple storyboard, either visually or narratively, to help the team envision how the proposed experience might unfold in practice.

The WebApp. To further enhance accessibility and address the logistical challenges of the physical format, a web-based version of the CoDesign Tool is another goal of the PERCEIVE project redesign process. While the paper version remains valuable for in-person sessions, the digital prototype was created to extend the tool's usability to hybrid and fully remote contexts, ensuring broader reach and flexibility.

Initially developed in Figma to support early design work and internal collaboration, the digital version is now being implemented using the open-source ATON framework, the technological backbone of the PERCEIVE platform. This transition maintains the collaborative and interactive logic of the original tool, while introducing several new features tailored to the digital environment:

- Remote collaboration: Teams can co-design from different locations, accessing the tool from any device with an internet connection.
- Integrated moderator role: A built-in configuration flow supports session setup, including user invitations and stage guidance.
- Live progress tracking: Completed stages are visually marked and stored, providing an overview of progress throughout the session.
- Expanded card interactions: Cards include extended descriptions, hover effects, and voting tools to support more nuanced collaboration.
- Seamless integration with other PERCEIVE tools: The digital CoDesign Tool connects with the Prototyper for 3D visualisation and the Experiment Planner for testing cognitive and emotional impact.

The final web application will be released by the end of the project and made openly available through the PERCEIVE platform, ensuring continuity and wider adoption across institutions, disciplines, and design contexts.

2.2 Prototyper: A Web3D Authoring Platform for Hybrid Museum Exhibitions

As cultural heritage experiences increasingly evolve into hybrid formats, combining physical environments with digital elements such as augmented and virtual reality, IoT systems, video mapping, and Web3D technologies, the need for flexible, modular, and accessible authoring tools becomes a key concern in the design of contemporary exhibitions. These tools must not only respond to the technical complexities of such systems but also support curators, designers, and cultural professionals in shaping meaningful experiences. This is especially relevant when considering the objectives of the PERCEIVE project, which places strong emphasis on designing emotionally and cognitively engaging experiences for museum visitors and the broader public. Central to this vision is the idea of fostering feelings such as Care and Authenticity through immersive storytelling and interactive participation. These experiences often require sophisticated embodied interactions, immersive 3D environments, and the ability to combine visual, auditory, and spatial stimuli in a coherent narrative flow.

To address this challenge, the Prototyper [29] has been developed as a web-based 3D authoring platform tailored to the needs of Hybrid Exhibition Design (HED). The tool enables users to configure, simulate, and iterate on complex exhibition scenarios that blend physical and digital components. It is designed not only as a standalone platform but also as a part of the PERCEIVE design toolkit, where it can be effectively combined with the Co-Design Tool to bridge collaborative conceptualization and immersive prototyping. Built on the open-source ATON Framework, the Prototyper adopts a layered authoring methodology structured around three key abstractions: components, widgets, and templates. This structure allows for the modular construction of exhibition scenes while supporting varying levels of user expertise. Through this system, curators, designers, and researchers (including those without programming knowledge) can collaboratively design and test hybrid setups in an intuitive and visual way.

The Prototyper MVP (Minimum Viable Product) includes a growing set of authoring features. Among the main widgets currently available:

- **3D Model Layer Widget**: Allows users to import, place, and organize 3D assets (artworks, furnishings, devices) within the scene. Assets can be positioned and scaled using a built-in gizmo interface.
- **Viewpoint Widget**: Enables the definition of fixed or dynamic camera positions to guide user navigation or simulate visitor perspectives across different narrative stages.
- **Semantic Node Widget**: Associates potential metadata or narrative content to specific objects or areas.
- **Measurement Widget**: Supports spatial validation by allowing users to measure distances between points or objects within the 3D environment.
- **Texture Editor Widget**: Allows real-time customization of graphic panels, screens, and labels; useful for configuring interpretive content or visual identity elements.
- **VR Preview Mode**: Enables immersive testing of the scene, giving designers and curators a first-person experience of the configured exhibition (Fig. 2).

The Prototyper has undergone early testing in classroom settings and during its public presentation at SITEM 2025, where 12 participants from curatorial and design backgrounds interacted with a demo template simulating a hybrid archaeological exhibition. While levels of technical expertise varied, all participants were able to engage with the tool's core features, confirming its usability and the potential of its no-code approach. These initial results suggest that the

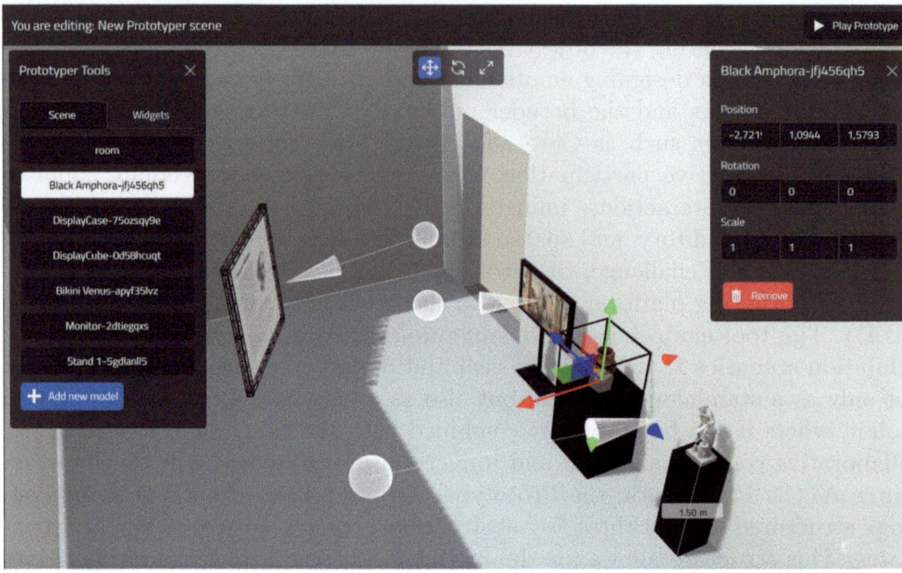

Fig. 2. Screenshot from a customisable scene in PERCEIVE Prototyper

Prototyper is on a promising path toward becoming an accessible and effective authoring environment for hybrid exhibition design.

Furthermore, the integration with the PERCEIVE Co-Design Tool allows for a complementary workflow: conceptual scenarios co-created in collaborative workshops (with cards and game-boards) can be rapidly transferred into the Prototyper as editable 3D prototypes, ready for simulation and refinement. This workflow bridges the gap between ideation and implementation, enhancing the collaborative and interdisciplinary nature of exhibition design.

The Prototyper has been actively used in the design process of the demonstrative prototypes for PERCEIVE's Final Exhibition, specifically in the installations focusing on Authenticity and Caring. Its integration into the design workflow has proven highly valuable, both as a practical tool to support and accelerate the prototyping of hybrid experiences, and as a means to reflect on and refine the authoring methodology itself.

Using the Prototyper throughout this process not only facilitated the iterative configuration of spatial layouts, roles, and multimedia content, but also allowed the team to identify new challenges, requirements, and technical directions for future development. The design process adopted a critical, hands-on approach, where the Prototyper was used to support the configuration of key elements. Its application also helped highlight areas that could benefit from further refinement or complementary strategies, such as the management of fragmented content, task balancing among participants, and the integration of content workflows, projection logic, sensor simulations, and narrative structuring. Overall, the experience provided both validation and insight—confirming the usefulness of the Prototyper in real-world hybrid design scenarios while generating valuable feedback to expand its feature set and improve interoperability. Aligned with the PERCEIVE project's vision of fostering empathy, participation, and cognitive resonance through digital cultural experiences, the Prototyper marks a significant step toward more inclusive, iterative, and sustainable practices in the digital transformation of cultural heritage interpretation.

2.3 Experiment Planner: An Evalutation Authoring Tool in Cultural Heritage Environments

The Experiment Planner is thought as a natural follow-up of the Prototyper, being conceived as a support for the creation of evaluation sessions for curators, museum practitioners and visitor scientists. Museological practice underscored the importance of the assessment methods in guiding the design and analysis of exhibitions: these scientific contributions even informed the creation of a novel branch of museum studies, i.e. *visitor studies*, which advocate for the "input from visitors [...] in the design of both exhibitions and programmes" and involves formal methodologies to "answering questions about the effectiveness and efficacy of exhibitions and programs" [27]. Additionally, exhibition assessment is not limited as an *ex post* analysis of a cultural initiative: scholars are increasingly calling for the wide implementation of *formative evaluations*, which "take place while an exhibition or program is still being planned or during the early stages of implementation. The results of such studies are intended to offer direct, concrete, and practical ways to improve a project" [48]. The Experiment Planner has its own theoretical foundations in this epistemological framework.

This tool ultimately allows to conduct visitor evaluations in 3D environments, in order to enable assessments also in early prototypal phase, without the constraints and the costs of physical set-ups. This approach is consistent with approaches in cognitive psychology, such as Unity add-ons like UXF [6] or bmlTUX [8]. Existing research however illustrates persistent gaps: in particular, most of these tools require technical skills (e.g. use of game engines) curators often lacks. Conversely, more user-friendly approach, such as Vizard and Navwell [13], provide limited flexibility and all these resources focus on psychologists as major target audience: similar endeavours for the digital heritage communities are still missing.

As a result, the Experiment Planner is thought as an authoring tools targeted to all those professionals (curators, museum practitioners, designers, but also psychologist and behavioural scientists) without specific coding skills to deploy experiments in virtual cultural heritage settings. In its current design, the platform wraps any ATON-based 3D environment (desktop or VR, also coming built through PERCEIVE prototyper) with an admin panel that guides these operators through the configuration and conduction of experimental sessions. It then enables the creation of evaluation sessions where participants can explore via VR headsets immersive environments (such as digital twins of heritage sites) and interact with digital replicas of exhibits. Visitors may also perform specific tasks (e.g., searching for an object) while the system streams in live to the admin position and state of the user, as defined in the evaluation protocol. At the end of the session, the tester receives a machine-readable report of all the testing sessions.

A first minimum viable product (MVP) was presented in a pilot study conducted on the digital twin of the exhibition *The Other Renaissance: Ulisse Aldrovandi and the Wonders of the World*, held in Bologna between 2022 and 2023 [4]. The prototype enabled psychologists and heritage experts to configure experimental conditions, assign tasks to participants, and observe their movements within a 3D environment. Nineteen experimental sessions were carried out with practitioners in psychology, demonstrating the feasibility of using digital replicas of exhibitions as reliable environments for cognitive studies. Besides some possible direction for future design iterations, the evaluation highlighted the strengths of the approach and confirmed the potential of the tool in bridging experimental psychology and cultural heritage, and in validating hypotheses on navigation, orientation, and appreciation of museum spaces (Fig. 3).

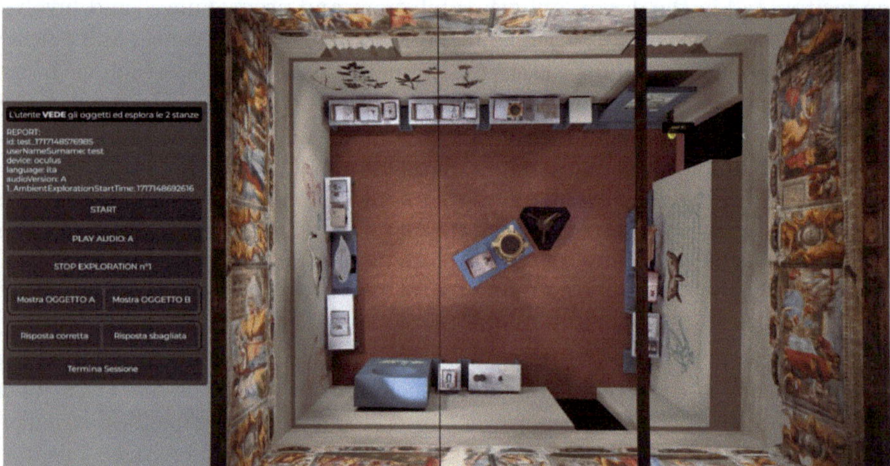

Fig. 3. Screenshot from a MVP of PERCEIVE Experiment Planner [28]

3 Prototypes

3.1 Authenticity

The concept of authenticity has long been central in heritage discourse, but often in restrictive terms: as the guarantee of material integrity or documentary truth. The Nara Document on Authenticity (1994) broadened this scope, emphasising cultural context and plurality of interpretation. Building on these debates, PERCEIVE explored authenticity not as an attribute of the reconstruction of cultural heritage objects or sites, but as a relational quality of the visitor experience [38].

Within this model, authenticity is articulated through three dimensions:

- Self: authenticity as personal resonance, when individuals feel that their encounter with heritage speaks to their own identity, memories, or aspirations.
- Others: authenticity as intersubjective recognition, emerging through shared experiences and dialogues within a community.
- World: authenticity as environmental and historical embeddedness, when heritage is perceived in relation to its broader social, symbolic, and ecological context.

The Gifts of Isis was conceived to activate all three dimensions simultaneously. Through its layered design, the installation allows participants to enter the reconstructed Iseum, embody historical roles, and engage in collective ritual practices that mirror, while also reinterpreting, those of antiquity. Authenticity is thus not given but produced through interaction.

The Gifts of Isis. This theoretical enquiry laid the foundation for the design of an *authenticity prototype* with Temple of Isis in Pompeii [14] as main case study: *The Gifts of Isis* (Fig. 4).

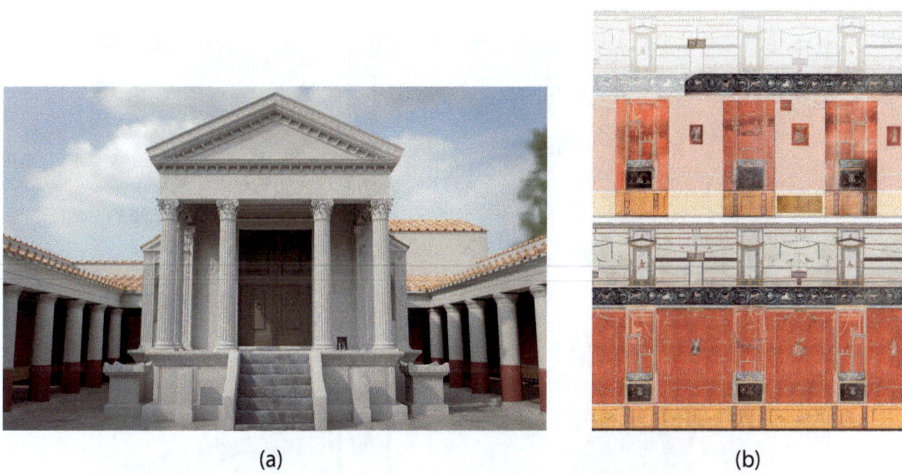

Fig. 4. Reconstruction of the Temple of Isis (a) and its frescoes decoration (b) (CNR - Institute of Heritage Science)

Concept and User Journey. The visitor's journey begins with a collective projection sequence. A narrator introduces the historical and religious significance of the Temple of Isis in Pompeii, highlighting its centrality in local devotion, its reconstruction after the earthquake of 62 CE, and its broader Mediterranean diffusion. Importantly, the narration does not simply present data but establishes a mood of reverence and curiosity. The closing words—"each fragment, if carefully observed, can still speak"—signal to participants that their role is not passive reception but active listening and questioning.

At this point, the three desks in the room become active, each linked to a digital device representing a symbolic gift to be requested from the goddess: Wisdom, Patience, or Creativity. By choosing a gift, each visitor also chooses a role within the narrative—Architect, Sculptor, or Painter. This selection establishes the Self dimension of authenticity: participants are encouraged to reflect on what they themselves seek, and how it resonates with personal experience.

The subsequent phase situates them in 62 CE, in the aftermath of the earthquake. The Popidii family has undertaken the reconstruction of the sanctuary, and each participant, through their chosen role, becomes part of this endeavour. Individual tasks follow: the Architect explores temple spaces through interactive plans, the Sculptor analyses statues of Isis to identify the most appropriate iconography, and the Painter reassembles frescoes from fragmented sources. These mini-games are more than diversions: they embody a process of learning through doing, trial and error, and discovery. Here, authenticity is produced not by receiving information but by engaging cognitively and playfully with fragments of the past (Fig. 5).

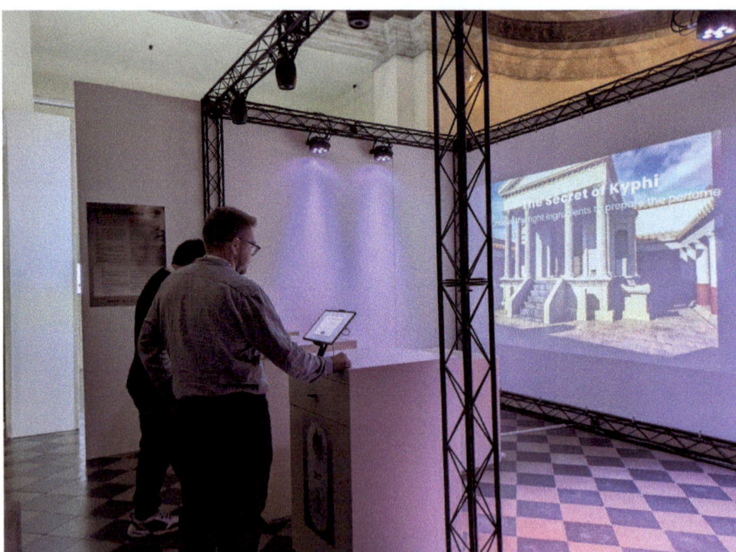

Fig. 5. Visitor interacting with the installation *The Gifts of Isis*

Having completed their tasks, the participants witness the reconstruction of the temple through a three-act projection: architectural spaces rising at dawn, sculptures populating the sanctuary at noon, and colours completing the environment by evening. This temporal progression mirrors the contributions of the Architect, Sculptor, and Painter, thereby affirming the value of each individual's labour. Yet at the same time, the narration shifts tone: Isis herself speaks, addressing each role with existential questions that dissolve the boundaries between past and present, work and selfhood.

This moment initiates the transformation of characters: the Architect becomes the Priestess, the Sculptor the Devotee, and the Painter the Musician. Participants, regardless of their own gender, are asked to embody both male and female roles during the experience, reflecting the historical composition of the Isiac cult while challenging stereotypical expectations. This metamorphosis also signals a shift in the dynamics of interaction. Whereas the first phase emphasised individual agency, the second centres on collective collaboration.

The ritual tasks exemplify this. Around the Priestess' desk, participants must select the correct linen robe by tactile exploration; around the Devotee's desk, they must identify sacred perfumes through scent; and around the Musician's desk, they must recognise the sistrum among other instruments. Each activity foregrounds a different sense—touch, smell, hearing—expanding the sensory horizon beyond vision and embedding participants bodily within the ritual space. At the same time, success requires dialogue, negotiation, and shared decision-making. This embodies the Others dimension of authenticity: heritage is experienced as a field of communal interaction rather than isolated contemplation.

The final sequence synthesizes the journey in a nocturnal procession, illuminated by lamplight, where all six characters—both from the reconstruction and ritual phases—converge. This collective procession symbolically links the ancient sanctuary with the present, as the voice of Isis returns the participants to contemporary concerns: conservation, reconstruction, and interpretation. By binding the ancient ritual to modern responsibilities, the narrative makes explicit the World dimension of authenticity: heritage is not an isolated relic but a living interlocutor with ongoing cultural, environmental, and political implications.

Importantly, this conclusion does not present authenticity as certainty or closure. Rather, it underscores that authenticity arises in the tension between evidence and imagination, between fragment and whole, between past and present. The visitor's mission ends not with possession of knowledge but with the recognition of heritage as a fragile, shared, and dynamic reality.

Installation and Set-Up. The exhibition *Perceive Isis' Colours* showcases two different prototypes: *Gifts of Isis* and *Echoes of Loss* (see *infra*). The entire exhibition can be mounted in a space of around 10×13 m. Placed within the half of the total exhibition area (in around $\mathbf{5 \times 5}$ **m**), *The Gifts of Isis* represents the second section of this exhibition. This immersive installation represents the collaborative and participatory phase of the narrative, where visitors collectively reactivate the symbolic and sensorial dimensions of the reconstructed temple.

- **Hardware:** one dedicated PC running the main control system (*VVVV,* https://vvvv.org/ *, and WebApp*); one local router; and three tablets running the participant web interfaces. The lighting system includes three white spotlights and two HEX RGBW multicolor lights mounted on the truss structure.

One projector is housed inside the central cabinet, aligned with the projection screen, and two loudspeakers provide ambient sound. A Bluetooth keyboard serves as an admin panel for the exhibition staff to start, stop, or manage sessions.

- **Furniture:** one truss cube structure (**3.4 × 3 × 3 m**) forming the main room framework. Three interactive desks, each with a tablet holder and a thematic drawer: (1) *Fabrics*, (2) *3D-printed musical instruments*, (3) *Scents*. One central cabinet contains the projector and audio equipment, and one projection screen is mounted on the back wall of the truss cube.
- **Graphics:** minimal environmental graphics and ambient lighting define the atmosphere of the ritual-like space. One *Illumiwall* vertical panel (2 × 1 m) describes the experience and contextualizes the interaction for visitors.
- **Installation:** assembly begins with positioning and securing the truss frame, followed by electrical wiring for lighting, audio, and projection. Network setup connects the control PC, router, and tablets, ensuring WebSocket communication between the *Room Controller* (VVVV app) and the *WebApp* interfaces. All software modules auto-start upon system boot, enabling synchronized operation between tablets and the immersive environment.

Evaluation Protocol. The goal of the evaluation protocol was to explore whether and how specific components from the Authenticity Framework were perceived by users as part of their experience within the exhibition space. The framework is structured across three dimensions—Self, Others, and World—each containing a set of components and corresponding elements, which were translated into design features during the co-design and prototyping phases. To qualitatively assess visitor perceptions, a custom questionnaire was developed, composed of 44 items distributed across the three dimensions of the framework. Each item was designed to reflect one or more components, using a Likert-type scale from 1 to 5 (1 = Strongly Disagree, 5 = Strongly Agree). The purpose was not to produce statistically generalizable results, but rather to capture subjective responses, identify patterns of perception, and understand which aspects of the experience resonated most strongly with visitors in terms of authenticity.

The structure of the questionnaire followed the framework's logic:

- *Self Dimension*: items addressed components such as Personal Disposition, Personal Context, Cognition, Perception, Emotions, Embodiment (personal), Familiarity (personal), and Time Evolution (personal). Participants reflected on curiosity, attention, emotional impact, mental and physical immersion, and perceived personal relevance or transformation.
- *Others Dimension*: items explored Language Exchange, Social Practices, Social Norms, Social Unpredictability, Social Embodiment, Social Familiarity, and Time Evolution (social), aiming to capture the role of interpersonal dynamics, collaborative tasks, group presence, and social resonance within the experience.

- *World Dimension*: this included components such as Verification, Physical Context, Atmosphere, Environmental Unpredictability, Environmental Embodiment, Environmental Familiarity, and Time Flow, focusing on how participants perceived environmental realism, coherence, sensorial atmosphere, and spatial engagement.

A final open-ended metric asked participants to rate how authentic they found the experience overall, on a scale from 1 to 10: "How would you rate how much you perceive this experience as authentic?".

The questionnaire also included basic personal data (age, education level, prior experience with Cultural Heritage and Technology, prior visit to the Iseum Temple in Pompeii, and previous interactive CH experiences). This contextual information supported a more nuanced qualitative interpretation of how background might shape the perception of authenticity. This protocol provided an exploratory, user-centered method to reflect on the practical application of the Authenticity Framework. It offered early insights into how visitors experience authenticity not only cognitively or emotionally, but also through sensory, social, and environmental dimensions—confirming the value of a multidimensional design approach.

Co-constructing Authenticity. By combining historical research, digital reconstruction, and multisensory participation, *The Gifts of Isis* articulates a new model of heritage engagement. It departs from traditional exhibition formats, where authenticity is guaranteed by the authority of curators or the aura of original artefacts. Instead, it proposes authenticity as something co-constructed through interaction—layered, provisional, and relational.

This dialogical model aligns with current reflections in museology and heritage studies that emphasise co-creation, inclusivity, and experiential authenticity. The PERCEIVE project contributes to this discourse by demonstrating how carefully designed game mechanics and narrative strategies can activate philosophical concepts in practice. Participants do not only learn about the Iseum; they experience its symbolic complexity, its ritual dimensions, and its relevance for their own lives.

3.2 Sense of Care

Narratives on cultural heritage often revolve around the concepts of conservation, preservation and ultimately care. Major contributions pertains to scientific protocols, also involving digital tools, for the custody of the cultural objects or practices: the understanding of care as a perspective to study the relationship between visitors and heritage is yet mainly unexplored. During its activities, PERCEIVE consortium tackled this major gap. Eventually, drawing on interdisciplinary literature review (moral philosophy, legal studies and curatorship), a first theoretical framework was elaborated [51]. As shown in Fig. 6, it conceptualises caring as a cyclical process composed of four core stages. Due to its circular structure, the act of caring has no definitive beginning or end.

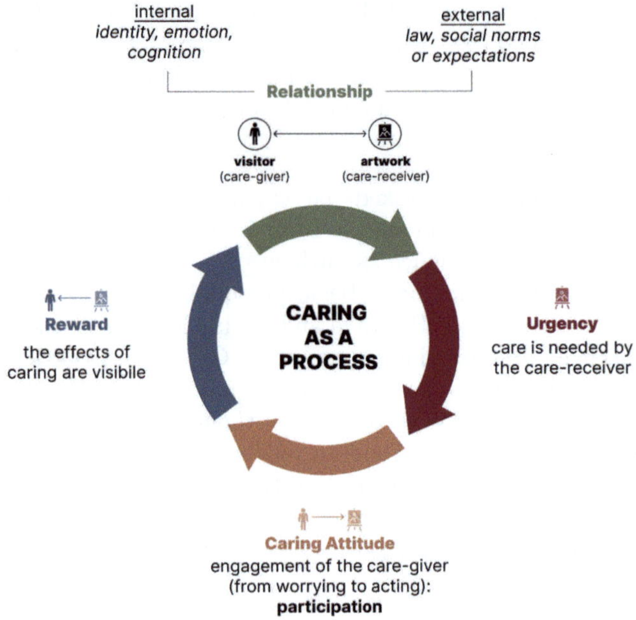

Fig. 6. Caring phases (or *components)* in the preliminary PERCEIVE Care Framework.

This model focuses first on the dynamic relationship between the caregiver and the care receiver, which can arise from internal motivations (e.g. emotional engagement, personal identification, or intellectual appreciation) or from external influences like legal obligations and social expectations. The second component pertains instead to the perceived urgency to act. The caregiver recognises that the care receiver (i.e. here a cultural object or site) is at risk and accepts a personal responsibility to respond to that need.

This perception of urgency gives rise to a range of caring behaviours or attitudes, which can vary widely in their purpose, effort, and degree of involvement. Within a museological context, these behaviours align with processes of audience engagement and participation. Finally, the framework highlights the phase of care-receiving, referred to as *reward*, in which the object of care responds to the attention it has received. For example, a restored artifact regains its original appearance or meaning, symbolising the successful fulfilment of caring needs. The outcomes of conservation and restoration work well exemplify this final stage of the cycle.

Echoes of Loss. This theoretical reflection on care was firstly applied to the phenomenon of lost polychromy in Greco-Roman sculpture. To this end, the design process drew upon an experimental inquiry on curiosity hooks in museum contexts: this analysis, conducted at the MANN, showed how audience consider the discovery of the scientific process itself as a truly engaging aspect [7]. These results informed the design of a second prototype, named *Echoes of loss*: as in *The Gifts of Isis*, the Temple of Isis in Pompeii is the main case study.

Concept and User Journey. This installation explores whether it is possible for interactive digital heritage experiences to foster visitors' sense of care, by providing access to the scientific colour reconstruction process as a curiosity hook. As a result, *Echoes of Loss* guides the visitor along an itinerary that mirrors the experience of care as described by Fig. 6: first, recognising loss and vulnerability; second, exploring the tools and methods of preservation; and finally, confronting the inherent uncertainty of reconstruction.

The first step of this itinerary is a short video (Fig. 7) that presents the splendour of the Temple of Isis at Pompeii, reconstructed through historical and scientific evidence, before confronting the devastation of its destruction—first by the earthquake of 62 CE, then by the eruption of Vesuvius in 79 CE. The visitors are invited to sit and wear headphones to watch and listen closely to the video, in a controlled space that aims to reduce distractions, thus allowing visitors to devote their full attention and time, fostering an emotional connection with what has been irretrievably lost.

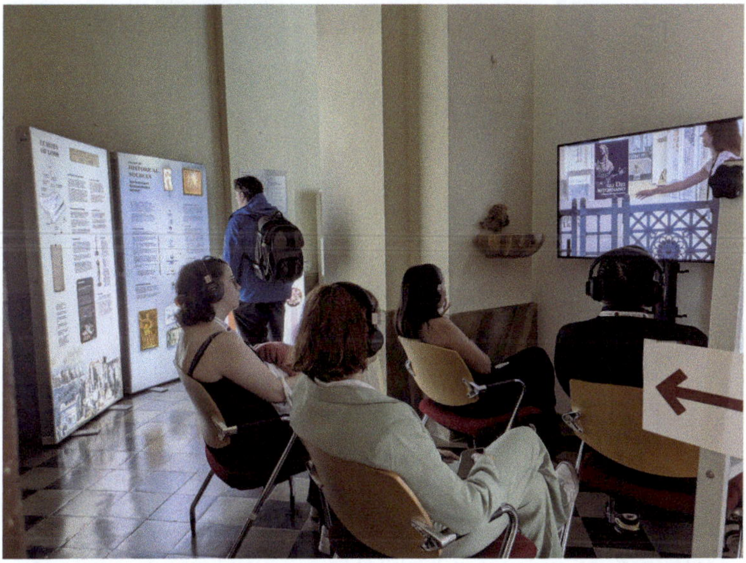

Fig. 7. Visitors watching Echoes of Loss video (Step 1)

The narrative moves between presence and absence, between the imagined vibrancy of colours and the starkness of archaeological remains. In doing so, it activates a sense of vulnerability: what is gone cannot return, but its traces still demand our attention. The affective tone is central here: the visitor is not merely informed but asked to feel the weight of loss. This moment reflects the first stage of caring: the recognition that something valuable might disappear, and that such disappearance matters.

From this emotional introduction, the visitor enters the *ColourLab* [Fig. 8]. Here, curiosity is channelled into active investigation. The laboratory space simulates the diagnostic processes used by researchers to study polychromy in classical statuary. Visitors manipulate replicas of scientific instruments, such as microscope, spectrometer, and special-filtered flashlight, placing them against a 3D-printed copy of the Anadyomene Venus (MANN, inv 6298), orginally located in the Temple. Its acquisition was characterized by a multi-modal approach combining photogrammetric acquisition (enriched with multiband data) [16] with high-resolution microscopy and localized spectroscopic analysis. Each gesture activates visualisations and explanations, showing how pigment traces are detected and analysed.

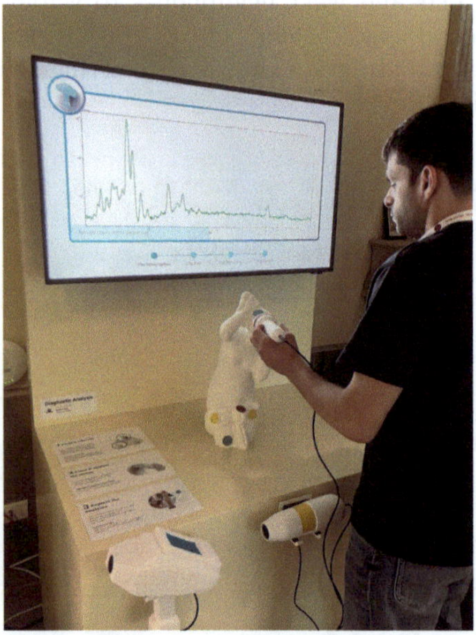

Fig. 8. Visitor interacting with the *ColourLab* installation (Step 2)

This participatory encounter highlights a crucial point: heritage science is not a distant process but one that can be made accessible and intelligible. By involving visitors in a simplified but authentic workflow, the exhibition demonstrates that care is not only the duty of experts. Instead, the act of learning about methods, of appreciating their complexity and limitations, is itself a form of care. Knowledge becomes a practice of responsibility.

The *ColourLab* culminates in a stage of synthesis: the transformation of fragmentary evidence into visual hypotheses. Here, visitors encounter mock-ups on marble surfaces, where pigments and binders have been experimentally applied, as well as digital simulations that map potential reconstructions onto the 3D model of the statue. This stage communicates an essential truth: there is no single "correct" reconstruction. Instead, scientific work generates a spectrum of plausible scenarios, each grounded in evidence yet open to interpretation. By exposing this provisionality, the exhibition reframes uncertainty not as failure but as part of a caring practice. To acknowledge fragility and indeterminacy is itself an act of respect toward the past.

The final installation, *Echoes of Colours* (see Fig. 9), brings these insights into an embodied form. Through an interactive projection mapping, the Venus Anadyomene is gradually "recoloured" as visitors lift a series of tangible objects, progressively heavier. Each object triggers a projection sequence, moving from chaotic possibilities to constrained hypotheses, and finally to a carefully reconstructed vision.

The increasing weight of the objects serves as a metaphor for the weight of the research: to engage with heritage requires effort, persistence, and care. This metaphor is not incidental. It conveys the principle that heritage care is not passive reception but an active, sometimes demanding practice. Visitors perceive, quite literally in their hands, that the recovered image is not definitive but provisional, the outcome of negotiation between evidence and imagination.

This moment can be seen as the "reward" phase of the care process: after the recognition of loss and the labour of research, visitors witness the temporary "healing" of the statue through colour and light. While this healing is symbolic rather than material, it creates a genuine relationship between person and object. The visitor has not preserved the statue physically—which remains the task of conservators—but has enacted care through attention, empathy, and intellectual engagement.

The three stages of *Echoes of Loss*—loss, investigation, and reconstruction—thus articulate a broader understanding of heritage care. Instead of equating care with conservation, the exhibition positions it as a multi-layered practice: emotional, cognitive, and participatory. This approach resonates with insights from philosophy and psychology, where care is understood as a relational and distributed activity rather than a specialised skill.

Installation and Set-Up. In the set-up of the exhibition *Perceive Isis' Colours*, *Echoes of Loss* makes up the first section and occupies approximately half of the overall area (around 8×7 m).

This area hosts a sequence of three interconnected installations that guide visitors through the theme of lost polychromy, moving from emotional reflection to scientific exploration and finally to immersive visualization. The section is composed of the following components:

Video Introduction. A self-standing 55-in monitor with integrated hooks for four Bluetooth headsets forms a small, enclosed viewing area where visitors can sit and watch a 3-min introductory video in loop. The setup includes:

- **Hardware:** 55-in display on an aluminium stand; 4 Bluetooth headphones; power connection and media player.

- **Furniture:** four chairs and, optionally, a partition element to create a more intimate area, visually separated from the rest of the exhibition path.
- **Graphics:** one *Illumiwall* vertical panel (2 × 1 m); minimal wayfinding graphics inviting visitors to watch the video.
- **Installation:** the monitor structure is self-supporting and easy to transport and mount; audio devices are hung on dedicated hooks positioned along the vertical pole of the self-standing structure. The *Illumiwall* panels are lightweight, self-standing aluminium frames with stretch-mounted printed fabric and backlight modules (25 cm depth). Each unit requires only a power connection.

ColourLab. An interactive module exploring the theme of polychromy and material investigation through 3 printed elements and tangible interactions.

- **Hardware:** Arduino-based system connected to a Mini PC managing tag recognition and video playback. Eight RFID tags are attached to the surface of a 3D-printed statue, each decorated by a circular colored sticker marking a diagnostic point. Three 3D-printed diagnostic tools (a *spectroscope*, a *portable microscope*, and a *UV flashlight*) act as readers, each associated with one color group. Each instrument reads only the tags of its corresponding color, triggering the playback of specific videos on the monitor placed above and behind the statue.
- **Furniture:** one cabinet supporting the statue and the tangible interface (the three diagnostic tools). The monitor is mounted on a vertical frame behind the statue. Two *Illumiwall* vertical panels (2 × 1 m) and one *Illumiwall* desk module (1 × 1 m) complete the station, providing information and color samples.
- **Graphics:** printed panels (*Illumiwalls*) show scientific content, color samples, and descriptions related to the diagnostic analysis' process.
- **Installation:** the *Illumiwall* units are plug-and-play and require only a power connection. The cabinet houses all electronic components (PC, Arduino, and sensors) and connects directly to the power supply. Upon activation, the system automatically launches the *ColourLab* application.

Echoes of Colours. An interactive installation combining video mapping, tangible interfaces, and a sculptural display, inviting visitors to engage physically with the projected reconstruction of the statue's lost colors.

- **Hardware:** two projectors (one integrated in the main cabinet and one inside the statue pedestal), one Mini PC managing the projection mapping and sensor input, and three tangible sensors shaped as liftable cylindrical weights. The sensors communicate directly with the PC through a custom Arduino board, triggering audiovisual transitions in real time.
- **Furniture:** one main cabinet integrating the tangible interface and projector; one pedestal supporting the 3D-printed statue and containing the second projector; one projection wall aligned precisely with the statue; one *Illumiwall* vertical panel (2 × 1 m) completing as introduction for the application.

- **Graphics:** the *Illumiwall* introduces the application describing its metaphore about the process of color reconstruction. The projection wall serves as a dynamic surface displaying light and pigment animations mapped onto the 3D statue, creating a layered interplay between scientific data and artistic expression.
- **Installation:** the projection alignment between the cabinet projector and the statue is the most critical step. A custom vertical mount allows fine adjustment *in situ* for perfect registration of the mapping. If the setup significative changes, the content can be recalibrated using video-mapping software and re-rendered accordingly. The entire module is pre-wired and modular; only power and video connections are required during setup.

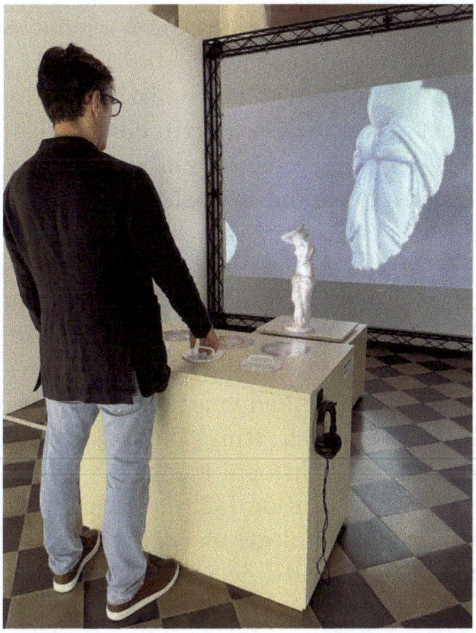

Fig. 9. Visitor interacting with the *Echoes of Colours* installation (Step 3)

Evaluation Protocol. The prototype was developed following the principles of *concept-based design* [49] and role of technologies as persuasive agents [18]: the core goal is indeed having visitors reflect on their own relationship with cultural heritage and understand how practices of care are accessible also to a non-specialist audience. To verify this assumption, the evaluation protocol is based on three core sections: (a) definition of a new metrics to assess caring attitude towards cultural heritage; (b) testing session with repeated measures to observe change in visitors' intentions; (c) use of tracking devices to collect advanced analytics of visitors interaction.

These needs faced as a major drawback the lack of existing evaluation resources specific to the sense of care in museum contexts. As a result, the construction of the protocol commenced with the definition of a new scale, the *scale for the assessment of caring attitude towards cultural heritage* (CHARE) [52] (a), in collaboration with Department of Psychology of the University of Rome "La Sapienza" and University of Bologna. This resource, which has been validated on a sample of young adults, consists of 16 different questions, which investigate real actions performed by the visitors over the last year. These items are ordered in increasing level of difficulty and enabled the construction of Rasch model, through which it is possible to measure *caring attitude* as a latent ability.

To better understand whether the effect of this prototype is significant only for those audiences with already high attention to cultural heritage preservation and valorization issues, CHARE metrics are used as a co-variate of the other items in the core part of the protocol, i.e. a test with repeated measures (b). In the pre-tests, data concerning anagraphics, professional/education background and responses to CHARE scale are collected. Additionally, a set of seven items is included concerning: intention to show three distinct caring behaviours; perception of individual, collective and participative efficacy [54]; support to cultural events or initiatives. These results are later matched in the post experience questionnaire, where these seven questions are asked anew. Furthermore, this second form includes further items concerning the different modules of the exhibition: the emotional response in the initial video; the knowledge acquisition in the *ColourLab* installation; the perceived meaningfulness, general appreciation and the activation according to SAM test [41] for the final projection.

Lastly, the final phase of the protocol involves the retrieval of advanced analytics concerning visitors' cognitive and emotional responses (c) through a tracking device: the BIOPAC Research Ring (https://www.biopac.com/product/research-ring/). Compared to traditional tracking systems, this ring provides a lightweight and fully wearable solution [42] for acquiring multiple physiological and behavioral signals, including electrodermal activity (EDA), photoplethysmography (PPG), finger temperature, accelerometry-based movement, and electrocardiogram (ECG) data. Furthermore, it is compatible with BIOPAC's Acknowledge Research Ring Software [2], enabling both real-time recording and cloud-based data import. This approach is supported by previous studies in which similar devices have been successfully employed to analyze visitor engagement in digital heritage contexts with promising result [15]. Accordingly, participants will be invited to wear the device throughout the entire experience, and the collected physiological data will be integrated with self-report measures from the post-experience questionnaire.

Heritage Care as Collective Endeavour and Responsibility. By inviting visitors to perceive heritage as fragile, to learn about the methods of its study, and to accept uncertainty as intrinsic, the exhibition reframes cultural heritage not as a static repository of objects but as a living field of responsibility. The past cannot be saved once and for all; it must be continually reimagined and cared for through collective engagement.

In *Echoes of Loss*, the exhibition proposes a new model of cultural heritage engagement: one in which visitors are not passive spectators but active carers. Through a carefully designed itinerary, the section translates theoretical reflections on care into embodied, multisensory experiences. It shows that caring for heritage involves recognising loss, engaging with knowledge, and embracing uncertainty—not as obstacles, but as vital components of a shared responsibility.

Ultimately, the sense of care cultivated here extends beyond the Temple of Isis and its statues. It calls visitors to consider how their everyday actions—whether learning, sharing, or advocating—participate in preserving the fragile threads that connect us to the past. In this way, the exhibition does not only revive the colours of Isis; it illuminates the possibility of a future where heritage care is a collective endeavour, rooted in empathy, awareness, and commitment.

Tiny Conservators. *Tiny Conservators* is an interactive caring prototype developed by the MUNCH Museum in Oslo as part of the PERCEIVE project. It will be installed in December 2025 in The Amfi, the museum's circular glass lobby. The project's primary goal is not only to communicate the science of colour change in artworks, but to cultivate a culture of care among children and families.

The case study focuses on Edvard Munch's The Scream (1910 version), a painting whose dramatic sky colours are highly vulnerable to degradation. Through playful cooperation, children learn that colours fade when exposed to light, humidity, and time. But the prototype goes further: by role-playing as guardians of colour, children also learn to practice care with each other, with the ambition of reinforcing empathy and collaboration as equally important conservation values.

The development of *Tiny Conservators* was carried out by an interdisciplinary team at MUNCH, including game designers, an art mediator, technical and physical light and exhibition designers, and a conservation scientist. This collective expertise allowed the prototype to integrate scientific accuracy with playful storytelling, digital immersion with tactile inclusivity, and art mediation and participatory design with children.

Cognitive Goals and Design Approach. The prototype's cognitive goal is to move children from abstract knowledge ('colours change over time') to lived awareness ('I can care for fragile things'). Its design is shaped by four principles:

- Care through role-taking: children become "Tiny Conservators" inside the microscopic layers of the painting, symbolically taking responsibility for fragile pigments.
- Narrative mediation of threats: the destructive forces—Time, Humidity, and Light—appear as flamboyant "Monarch of Decay."
- Care through cooperation: the game supports 1–6 players, who must collaborate to shield colours and protect each other's avatars.
- Participatory design with children: primary school students were involved in co-design workshops, shaping character concepts and mechanics.

Case Study and Storyworld. The storyworld begins in a hidden, magical realm within the paint layers of The Scream. Here live the Tiny Conservators, miniature guardians devoted to protecting colour. Their adversaries are the Monarhcs of Decay, whose powers embody the fragility of pigments: dazzling Light burns, slippery Humidity erodes and accelerating Time speeds decay. Through this narrative, children learn that artworks—like living beings—are vulnerable and require care. By overcoming the Monarchs of Decay together, they practice empathy not only toward the artwork but also toward each other: protecting teammates, sharing resources, and celebrating victories as a group (Fig. 10).

Installation and Interaction Design. Opening in December 2025, the installation will transform *the Amfi* into a participatory conservation arena. The Amfi is a circular, glass-walled lobby space in the MUNCH Museum, and its distinctive architecture makes it an ideal setting for a communal and immersive experience. Families and school groups are invited into an amphitheatre-style circular room, ensuring that play is both collective and visible to all (Fig. 11).

The physical environment is integral to the prototype's design. Large-scale projection and wrap around glass walls, while responsive coloured lighting floods the space in reaction to in-game events. Surround sound further amplifies the drama of the Royal Elementals, creating a multi-sensory encounter that extends the digital storyworld into the physical room.

At the heart of the installation are the "conservosauruses" game stations—tactile, child-friendly controllers designed with inclusive access in mind. Their form reflects their digital counterparts and are adapted for young players, with ergonomic grips and simplified inputs that ensure children of varying abilities can participate. The conservosauruses anchor the digital play in a tactile and embodied interaction, reinforcing the project's aim to make conservation visible, touchable, and emotionally engaging (Fig. 12).

Immersive Environments and Child Engagement. The design of Tiny Conservators builds on research showing that immersive environments can enhance children's learning and emotional engagement. Immersion is not only a technical quality of space but also a social and affective one: it draws participants into a world where sensory, cognitive, and emotional responses converge.

Studies of multimodal play environments have demonstrated how immersive computing environments support children's engagement through physical movement, social collaboration, and embodied interaction [3]. Similarly, observational studies in museum education have shown that immersive art spaces can foster curiosity, emotional involvement, peer collaboration, and reflective thinking among children aged 9–11 [25]. These findings indicate that immersion is not merely an aesthetic feature but an important factor in supporting children's ability to connect with abstract or fragile concepts.

Game-based learning in museums has also highlighted the role of immersion in creating meaningful and memorable experiences. By embedding game mechanics into the museum environment, immersive designs promote deeper engagement and sustain visitors' attention [24]. Research on children's voices in

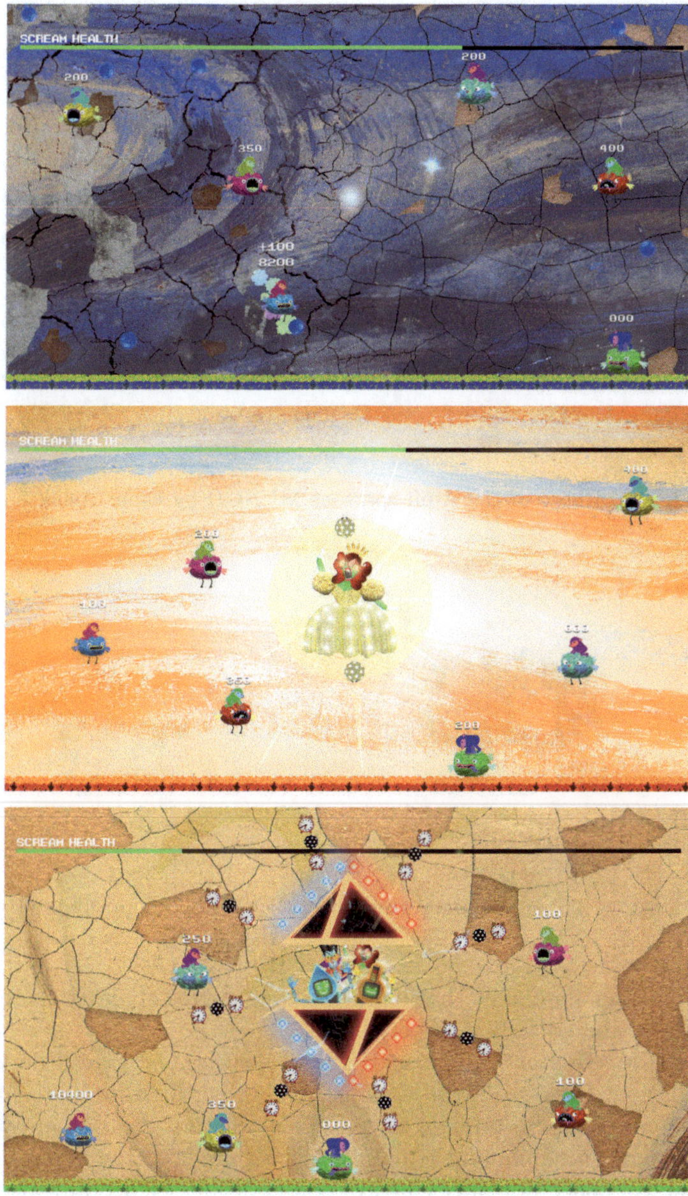

Fig. 10. Images from *Tiny Conservators*: (i) Encounter with Royal Elementals representing Time, Humidity, and Light; (ii) Cooperative play where children defend pigment health together; (iii) Cooperative strategies against elemental forces to preserve the artwork

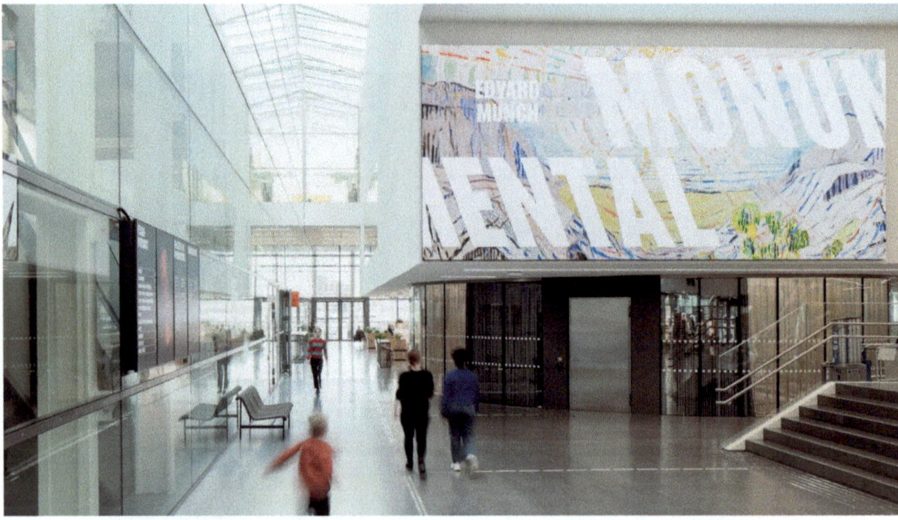

Fig. 11. Photo of the Amfi space at MUNCH by Einar Aslaksen

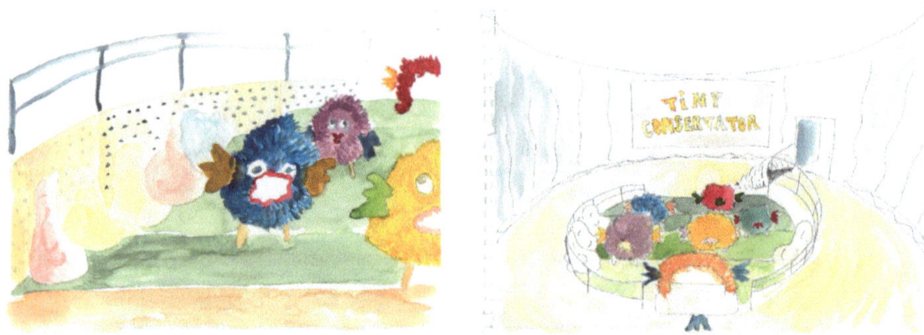

Fig. 12. Illustration of Tiny Conservators in Amfi at MUNCH, by Nynne Ryhl Olsson.

digital play further reinforces that children value environments that feel responsive, embodied, and socially connected [26].

For *Tiny Conservators*, this research base justifies the hybrid design of The Amfi installation: large-scale projections, responsive lighting, surround sound, and tactile conservosaurs combine to create an affective environment where conservation becomes tangible and felt. In line with previous findings, the installation's immersive space not only conveys conservation science but also supports social care behaviours such as helping peers, protecting avatars, and negotiating shared strategies.

Evaluation Protocol. The evaluation of *Tiny Conservators* is designed to be pragmatic, anonymous, and group-based, with emphasis on collective experience rather than individual data. It builds on a participatory design (PD) approach, which is recognised as a valuable method for engaging children in ideation processes and co-creation of interactive experiences [22]. PD is not only about generating content with children but also about creating conditions for mutual learning: children contribute their creative perspectives, while adult design partners gain insight into challenges as seen from a child's point of view, resulting in a deeper understanding of how experiences are received and played with [30]. The evaluation relies on three simple, non-intrusive methods:

1. *Observation*: documenting engagement, cooperation, and care behaviours such as protecting peers' avatars, negotiating strategies, or celebrating shared successes.
2. *Teacher feedback*: short, informal conversations to understand how children interpreted and emotionally connected with themes of colour change and fragility.
3. *Group reflection*: informal discussions with the class, capturing collective impressions while maintaining full anonymity.

Workshops & Participatory Design Activities. Between January and June 2025, five workshops and playtests were carried out across 5 workshops with schoolchildren, across grades 2–7. These sessions were not formal research studies but part of developing the prototype through participatory design. They served to test concepts, refine mechanics, and integrate children's own ideas into the narrative and character design:

- January ($\approx$20 children, grade 2): Focus group exploring colours, materials, and prior knowledge of *The Scream.*
- February ($\approx$25 children, grade 2): Workshop refining conservation concepts and story ideas.
- March ($\approx$25–30 children, grades 2 & 4): Co-design session where younger children invented the "Tiny Conservators" characters and older children developed the villains (Time, Humidity, Light).
- June ($\approx$20 children, grade 7): Playtest with older children to assess difficulty, pacing, and engagement.
- June ($\approx$15 children, grades 2–4 mixed): Playtest focusing on cooperative strategies and group dynamics.

Through these workshops, children were not only participants but also co-creators of the experience. Their contributions ensured that the game mechanics, storyworld, and interaction design reflected how they themselves think, play, and care. The final stage of evaluation will take place during the December 2025 installation at MUNCH, again using observation, teacher dialogue, and group reflection. This approach ensures that evaluation remains anonymous, ethical, and participatory, aligning with the project's aim to foster both knowledge and care.

Caring and Empathy through Play and Participation. As a caring prototype, *Tiny Conservators* demonstrates how conservation science can be reimagined not only as knowledge transfer but as a practice of empathy. By involving children as co-designers and players, the project demonstrates how museums can foster habits of care—care for cultural heritage, care for peers, and care for the fragile things that matter in everyday life.

Research on serious games has shown their potential to complement social and emotional learning [53] and to enhance children's emotional intelligence through playful exploration of affective scenarios [33]. This project builds on such findings by extending empathy from social relationships to the domain of cultural heritage. By protecting colours together, children learn that artworks, like living beings, are vulnerable and require care. In doing so, the prototype not only communicates conservation science but also cultivates emotional and social engagement—an outcome increasingly recognised as vital in heritage education.

The modular design of *Tiny Conservators* ensures that the approach can be adapted to other artworks, while the open-source codebase supports replication by partner museums. Most importantly, it demonstrates that games can generate empathy not only toward people but also toward fragile cultural objects, reframing conservation as a shared, participatory act of care.

3.3 Participation

The third axes of PERCEIVE consortium's research on digital heritage experience revolves around the concept of *participation* [46]. Its main outcome is the Open Space Museum (OSM), which presents new strategies to extend civic participation in the digital domain.

Open Space Museum. The Open Space Museum (OSM) is a prototype of a mobile cultural infrastructure designed to extend museum experiences beyond institutional walls [11,12]. Developed within the framework of the PERCEIVE project, the OSM seeks to foster public participation, accessibility, and dialogue through a modular and transportable exhibition architecture. The Basel work phase represented the first opportunity to assemble and evaluate the full OSM structure under real conditions. Conducted over several months in a large harbour hall, the experiment aimed to test both the mechanical systems of the structure and the visitor experience it affords. Beyond confirming structural stability, the work focused on refining workflows, ergonomic comfort, and spatial circulation—key dimensions for the future touring of the OSM across European contexts (Fig. 13).

Basel Work Phase Installation. The Basel work phase was carried out over an twelve-week period in early 2025. The testing combined practical construction, design evaluation, and observational research. The hall's uneven concrete floor created conditions approximating those found in outdoor or urban contexts, providing an ideal test environment.

A standardized workflow was developed to ensure repeatable construction and dismantling. The process began with the alignment of four base cubes

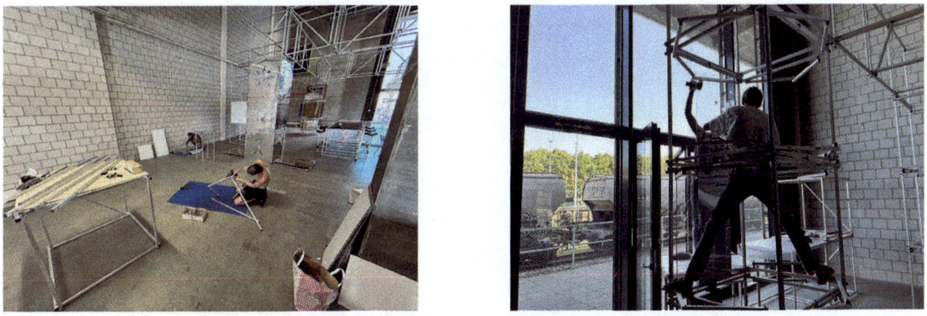

(a) (b)

Fig. 13. OSM structure under assembly during the Basel work phase, showing how pre-constructed cube elements are joined by four hands to create the matrix.

using custom 3D-printed shims to compensate for floor irregularities. Once leveled, Case Study Elements (CSEs) and Social Elements (SEs) were sequentially installed, followed by diagonal struts to stabilize the framework. The "ceiling" grid of cubes—housing integrated skylight modules—was installed last. The entire structure was subsequently dismantled and packed into five custom-built crates (1.1 m^3 each) designed for repeated transport and long-term storage.

Based on empirical observation, a three-person team is sufficient to assemble the OSM within five working days, with dismantling requiring only three days. This lean labor model demonstrates the scalability of the OSM for temporary installations and urban deployments (Fig. 14).

Basel Work Phase Outcomes. The Basel work phase evaluation highlighted the following features of the Open Space Museum:

Structural Performance. The introduction of 3D-printed shims and diagonal bracing eliminated structural racking and ensured precise leveling, resulting in stable alignment between the vertical and horizontal cube matrices. These refinements established repeatable construction tolerances critical for mobile deployment.

Spatial Design, Accessibility, and Ergonomics. The Basel test confirmed the importance of accessibility in shaping user experience. Pathways were widened to support wheelchairs and groups, while interactive components were re-positioned to be usable from seated or standing postures. Seating and rest areas within the SEs were optimized to create an alternating rhythm between engagement and repose (Fig. 15). This emphasis on universal design contributes to the OSM's role as an inclusive public environment rather than a conventional exhibition container.

Evaluation of Cultural and Social Elements. The Case Study Elements, containing cultural and scientific content, were reviewed for clarity, durability, and integration within the overall layout. Adjustments enhanced visitor flow and inter-

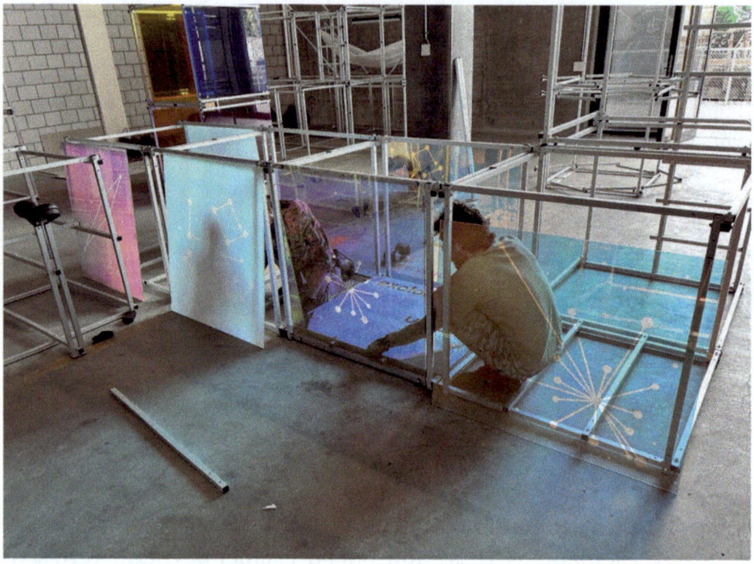

Fig. 14. Montage of the colored plates onto the aluminum frame structure. Here using dichroic film for the creation of the *Space Tunnel* Element.

pretive coherence. The Social Elements were analyzed for ergonomics and social facilitation. While most performed well, the *Lounge Chair* requires redesign due to construction inaccuracies, and the *Bike Seat Table* demands improved seating comfort. *The Hammocks* were successful but will benefit from short instructional videos (accessible via QR code) to guide novice users.

Among the Cultural Elements, the *Space Tunnel* and *DanceSquared* installations provided distinct forms of visitor engagement. The *Space Tunnel* offered an immersive sensory experience, surrounding visitors with mirrored, dichroic surfaces that transformed light and reflection into a spatial encounter with color. In contrast, *DanceSquared* introduced a participatory and performative mode of interaction, inviting visitors—especially younger audiences—to activate sound and rhythm via QR codes on their mobile devices. Together, these elements highlight the range of experiential modes within the OSM, from contemplative observation to embodied participation.

Modularity and Mobility. The Basel work phase demonstrated the OSM's modular and mobile design. Each cube and connector follows a standardized dimension and assembly logic, allowing the structure to adapt to varied architectural or urban contexts. This modularity enables flexible layouts and curatorial configurations, ensuring that the OSM can be reshaped for different sites and themes.

Equally important is its mobility. Although the installation occupied more than 600 m^2 fully assembled, it can be collapsed into five transport crates measuring $1100 \times 1000 \times 1000$ mm each (Fig. 16). The crates are engineered for rapid packing and movement, easily handled with a simple hand pallet truck. Together,

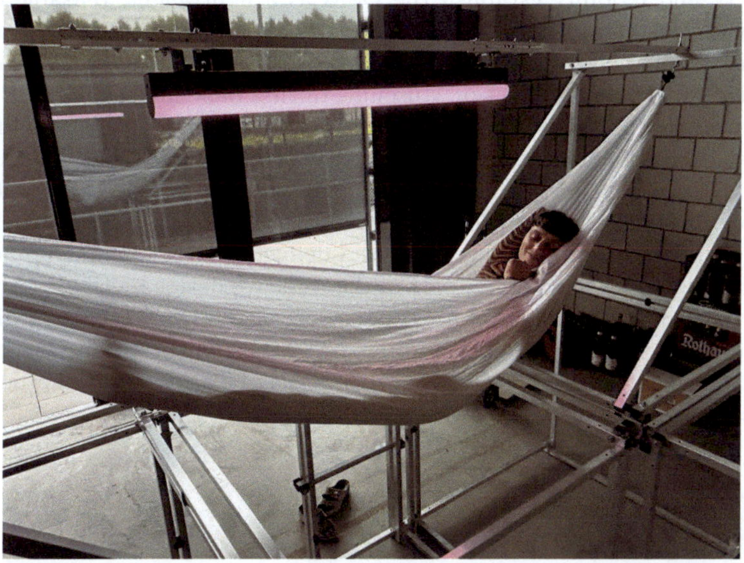

Fig. 15. Visitor interaction with a Social Element—*The Hammock*, designed to encourage rest, reflection, and social interaction within the OSM. ©Virtuale Switzerland 2025.

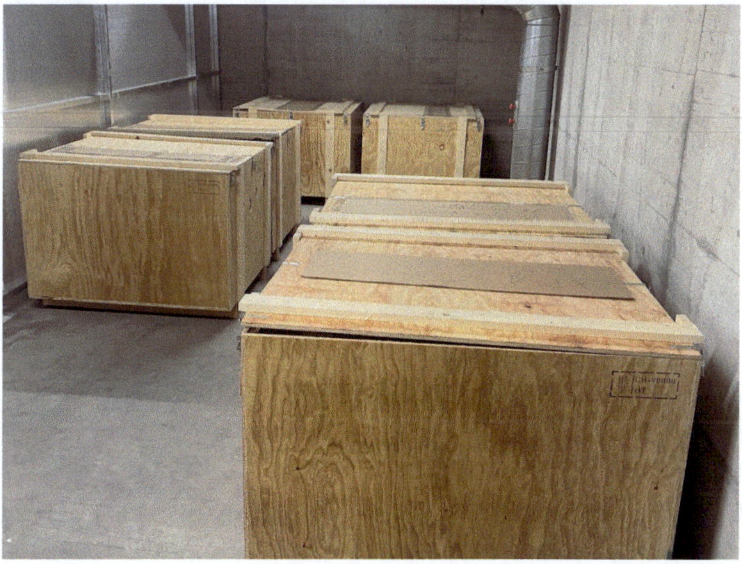

Fig. 16. The OSM crate system, showing how over 600 m^2 of exhibition structure can be collapsed into five transportable crates (1100 × 1000 × 1000 mm). Built for efficiency, the crates can be easily opened, closed, and moved with a simple hand pallet truck.

modularity and mobility establish the OSM as a lightweight, reusable cultural infrastructure that challenges the permanence of traditional museums and supports sustainable, site-responsive forms of exhibition-making and is an ideal alternative to the closed doors of Museums during pandemics [20].

In synthesis, the Basel phase validated the technical and conceptual feasibility of the OSM. The results highlight three key findings: (1) structural reliability, (2) operational efficiency, and (3) inclusive design. Nevertheless, several design challenges persist, including the replacement of faulty media, the further development of the digital realm, delayed media delivery and ergonomic refinements. These ongoing adjustments form part of the iterative design methodology underpinning the OSM. From a theoretical perspective, the OSM advances discourse on participatory museology and modular design for cultural infrastructures. It demonstrates how experimental museum forms can balance material pragmatism with social engagement (Fig. 17).

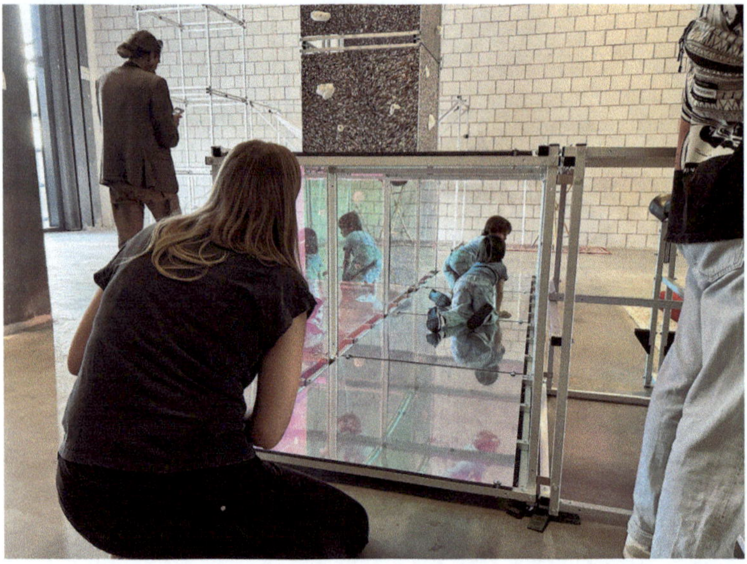

Fig. 17. Visitors inside the *Space Tunnel*—a mirrored, colored environment within the OSM that immerses participants in light and spatial reflection, creating a sense of being 'inside color' or 'in space'.

The Cultural and Social Elements together embody this balance. While Social Elements such as *The Hammock* emphasize inclusivity, rest, and social interaction, Cultural Elements like the *Space Tunnel* and *DanceSquared* expand the perceptual and performative dimensions of visitor experience. The *Space Tunnel* immerses participants in colour and reflection, prompting sensory awareness of light and movement, whereas *DanceSquared* transforms collective motion and sound into a participatory cultural performance. These complementary modalities demonstrate how the OSM merges architectural modularity with embodied interaction to create a holistic museum experience [9] (Fig. 18).

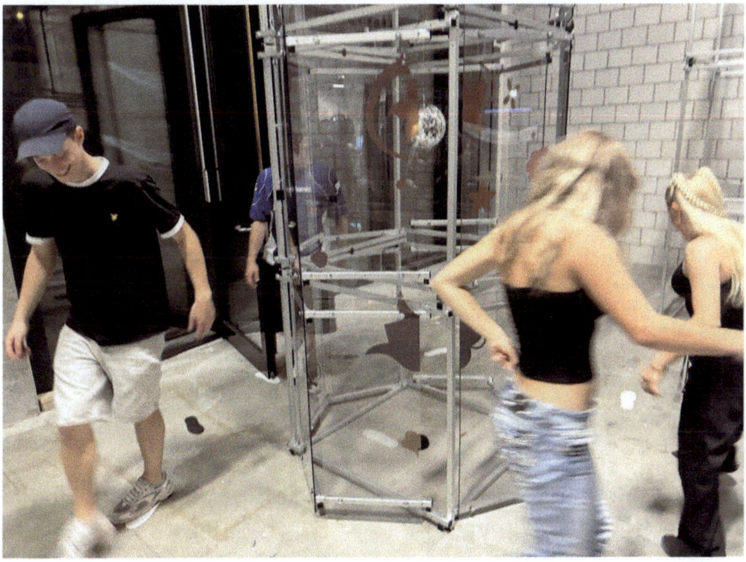

Fig. 18. Young visitors dancing around the *DanceSquared* Cultural Element, where QR-activated music turns the museum into a participatory sound space.

Phygitality: Hybrid Integration and the Emergence of Onlife Space. An essential dimension revealed during the Basel testing phase was the transformation of the visitor from passive observer to active participant within the Open Space Museum's phygital ecology [17]. The structure does not simply host digital content; it responds to presence, gesture, and perception. Each visitor becomes a sensor, interpreting variations in light, colour, material, and shadow as part of a responsive spatial dialogue. The CSEs including the multicolored skylights amplify this sensory awareness, while the QR codes positioned on each frame extend that perception into the digital layer, connecting the physical experience to web-based information, engagement apps, and augmented works.

This process of perception extends naturally into performance. The act of interacting, reacting, or navigating between illuminated structures constitutes a choreography of participation that activates the museum itself [23]. Movement determines perspective, light conditions, and the sequence of discovery. In this sense, the OSM operates as an open stage where the visitor's physical motion completes the spatial composition. The environment and its technologies are deliberately passive—there are no reactive screens or moving parts—but they are designed to invite action, using directed lighting and spatial rhythm to guide behaviour. The visitor's bodily presence gives the installation its temporal and experiential dimension (Fig. 19).

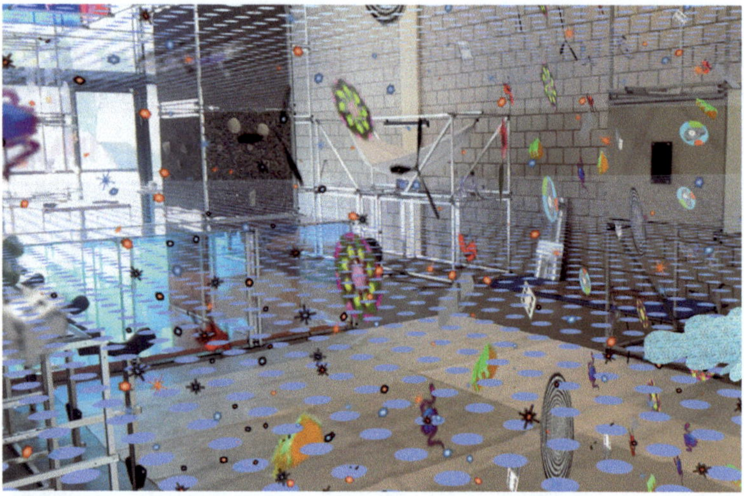

Fig. 19. AR artwork overlay displayed through the Hoverlay browser interface within the Open Space Museum environment.

Future Directions. The real-world testing of the Open Space Museum provided essential empirical data for its future deployment. The work established technical standards, refined design elements, and demonstrated that a lightweight, modular museum can operate efficiently within real environments.

The next phase will focus on refining the experiential components, including interactive media, engagement applications, and the further cultivation of digital twins, while expanding collaborations with cultural and research partners to explore the connection between art and science through creative inquiry. The Basel phase not only tested the feasibility of a structure but also articulated a conceptual and operational blueprint for a future in which museums extend beyond fixed locations to become dynamic, living systems of cultural exchange.

4 Demonstrators

In the PERCEIVE project, Demonstrators serve as functional research models that transform theoretical concepts about color, perception, authenticity, and care into tangible experiences. Their purpose is to make complex research accessible—bridging scientific investigation and artistic practice while engaging both expert and public audiences [34]. Conceived as compact case studies, they embody the project's interdisciplinary approach, linking technological innovation with curatorial experimentation.

4.1 Demonstrators as a Foundations for the Open Space Museum

The development of PERCEIVE demonstrators followed a practice-based approach grounded in constructing knowledge through making. This process focused on understanding how two-dimensional planes could be used to generate three-dimensional structures, a principle that later became the defining design logic of the OSM [32]. By manipulating form through planar construction, the team could test how colour, material, light, and spatial interaction combine that create wonderment and embody authenticity [39] and evoke a sense of care within each PERCEIVE scenario (https://perceive-horizon.eu/index.php/scenarios/) addressed.

Experimenting with transparent and coloured materials allowed for the study of how light can synergize multiple colour planes into new visual phenomena, producing effects that transcend the sum of their parts. These experiments not only explored the perceptual behavior of light and color but also demonstrated how architectural form could emerge from their interaction, and how the museum's spatial integration within its environment depended on the dispersion and blending of light in its surroundings. In this way, making became a form of research—revealing how material, color, and illumination can co-construct meaning, perception, and emotional engagement within hybrid museum environments.

Demonstrators were ultimately conceived as practice-based research tools to translate PERCEIVE's scenarios into tangible, multisensory experiences. By uniting scientific inquiry with artistic experimentation, they allowed color and perception to be studied through direct engagement rather than abstract data.

Each demonstrator functioned as both an experimental model and a communicative object, testing how color, authenticity, and care could be perceived, interpreted, and experienced across different media—from material to digital form. Through this approach, theory was transformed into practice, giving form to complex ideas about how light and color shape understanding, and in end effect offering a possible approach to the design of the Open Space Museum (OSM) [35].

The following section presents three key demonstrators that exemplify this process: the *Autochrome Demonstrator*, the *Scream Scaper Demonstrator*, the *VR CHROMA Demonstrator* and the *Gabinetto Segreto.*

Translating Research Into Experience. The PERCEIVE Demonstrators convert theoretical questions about color, perception, authenticity, and care into tangible, multisensory artifacts. Each functions simultaneously as an experimental model and a communicative medium—testing how material, light, and digital mediation can foster both understanding and empathy toward cultural heritage.

Together, the Demonstrators form a continuum of inquiry that spans multiple scales and modes of perception. The *Autochrome Demonstrator* (Fig. 20 explores the optical and AI-assisted reconstruction of historical color processes, using artificial intelligence to recover missing data and re-color degraded photographic imagery while maintaining transparency about its interpretive nature [36]. The *Scream Scaper Demonstrator* translates chromatic and emotional transformation into sound and light, reversing the role of the viewer by turning them into a color scientist who searches for areas of pigment loss and perceives the consequences of fading through multisensory feedback.

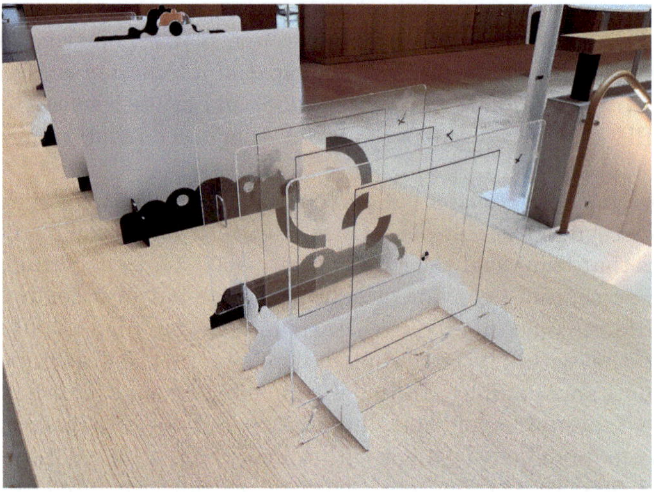

Fig. 20. The Autochrome Demonstrator—optical and AI-assisted reconstruction of historical color processes.

The *VR CHROMA Demonstrator* extends these experiments into immersive, virtual environments, where light and colour define both spatial composition and perceptual depth. Originally conceived as a comparative study between real-world and virtual colour perception, it provided the empirical basis for understanding how physical and digital color interact—knowledge that became central to the OSM's principle of phygitality. In addition, *VR CHROMA* offers a unique method to store and re-experience vintage AR artworks within a digital environment where they can once again be encountered and studied.

The *Gabinetto Segreto*, conceived as a portable micro-model of the Open Space Museum, condenses these ideas into a compact, luminous form that demonstrates how light itself can function as narrative structure. Taken as a whole, the Demonstrators establish a body of practice that links scientific analysis, AI-based colour reconstruction, artistic interpretation, and participatory care—collectively shaping the design philosophy of the Open Space Museum (OSM), where light, material, and perception coalesce into a living architecture of care.

The *Scream Scaper Demonstrator*. The *Scream Scaper Demonstrator* (Fig. 21), developed under the Fading Colors in Paintings scenario (Scenario 2) in collaboration with the MUNCH Museum, extends the study of color into the sensory, emotional, and temporal domains. Composed of three modular panels (each approximately $50 \times 51 \times 71$ cm) built from acrylic glass, it integrates laser distance sensors, an Arduino microcontroller, custom sound-synthesis software written in Processing, and synchronized LED lighting.

The demonstrator compares two photographic versions of Edvard Munch's *The Scream* (1938 and 1971) to visualize chromatic shifts caused by pigment degradation. Spectral data extracted from both paintings were mapped into a three-dimensional topography where height corresponds to hue difference. As the sensors scan the surface, the controller translates spatial data into tonal frequencies and light intensity, producing a live audio-visual composition. Visitors hear fading color as sound and see it as rhythmically pulsing light—a synesthetic translation of time into perception.

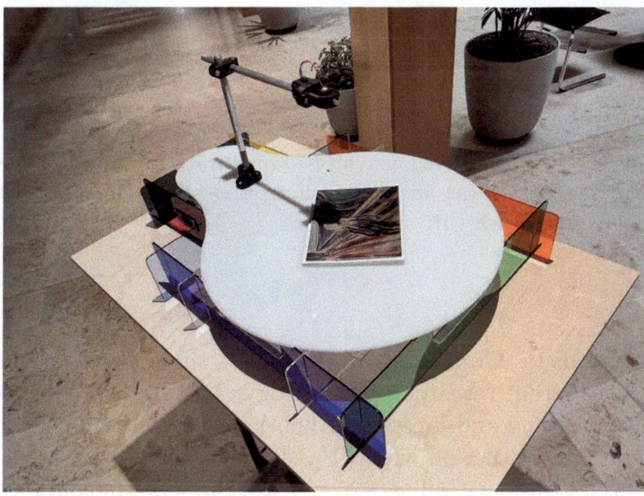

Fig. 21. The *Scream Scaper Demonstrator*—translating chromatic and emotional transformation into sound and light.

Through this real-time mapping, the demonstrator reverses the conventional viewing relationship: participants assume the role of color scientists, exploring areas of pigment loss and interpreting chromatic change as both data and experience. The installation reveals color as a dynamic and emotional event rather than a static attribute, transforming color loss into an act of expressive care.

The *VR CHROMA Demonstrator*. The *VR CHROMA Demonstrator* (Fig. 22) expands the inquiry into immersive digital space, operating within the Born Digital Art scenario and developed by the Immersive Realities Research Lab (IRRL) at HSLU, under the guidance of Virtuale Switzerland.

It consists of a VR headset ($15 \times 7 \times 20$ cm) mounted on a purpose-built display stand (approximately $120 \times 40 \times \times 40$ cm). The software, built in Unity 3D (C#) with AI-enhanced textures generated via StyleGAN, employs real-time global illumination to simulate changing light conditions.

Within the virtual environment, users explore a reconstruction of the Open Space Museum, complete with its Chroma Tower, AR Portals, and three integrated digital artworks—*Propaganda Stands* and Orators (John Craig Freeman), *Crystal Coffin* (Lily and Honglei), and *Space Art Gallery* (Arthur Clay).

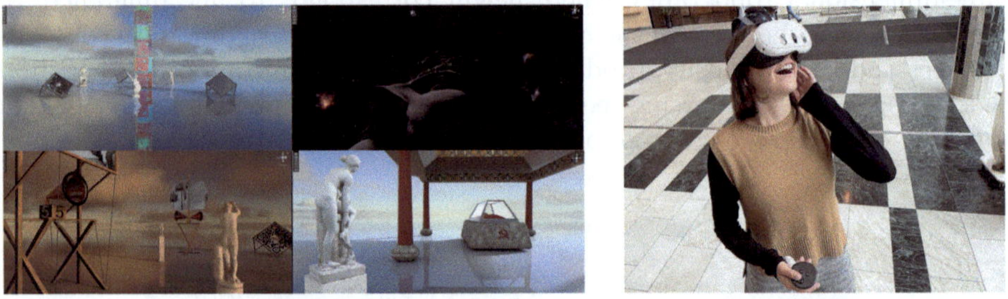

(a) (b)

Fig. 22. The VR CHROMA Demonstrator—comparative study between real-world and virtual color perception (a) and screenshot on visual comparison between digital and physical color experience (b).

Demonstrators as a Continuum. Together, these three principal demonstrators form a coherent research methodology. The *Autochrome Demonstrator* establishes transparency and layered authenticity through physical optics, combining historical reconstruction with AI-assisted recoloring to test how interpretation can remain visible within acts of restoration. The *Scream Scaper Demonstrator* introduces temporal and emotional mapping via real-time sensing, transforming the visitor into a color scientist who investigates pigment loss and perceives color change as both data and affective experience. The *VR CHROMA Demonstrator* extends these insights into immersive digital environments that fuse virtual and physical perception.

Among the later works, the *Gabinetto Segreto* (Fig. 23) plays a pivotal role as a micro-model of the Open Space Museum (OSM). Conceived as a portable, light-based environment, it condenses the OSM's principles into a compact, transportable form often presented at conferences to introduce the museum concept. Its projection system functions like a compact stop-motion film projector: laser-cut acrylic glass strips, engraved with imagery, serve as the film medium, passing through a guided light source to cast animated sequences of color and form onto surrounding surfaces.

This mechanism turns illumination itself into an act of storytelling, reinterpreting historical sensuality through the language of optical technology. Beyond its aesthetic value, the Gabinetto Segreto acts as a communicative prototype—a miniature, dynamic embodiment of how the OSM uses light to create participatory and emotional engagement.

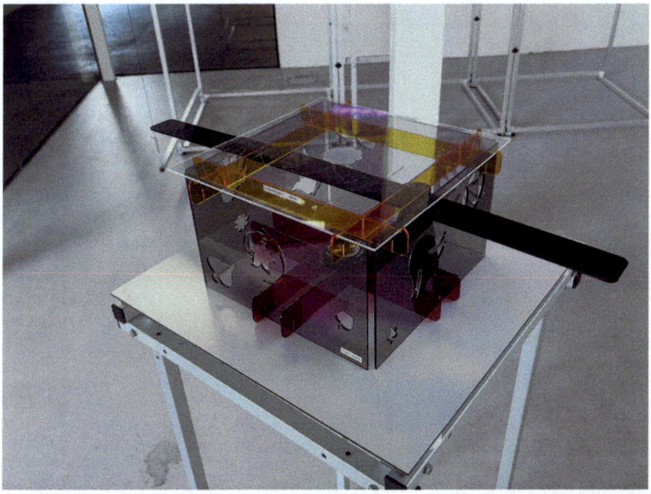

Fig. 23. The *Gabinetto Segreto*—portable micro-model of the Open Space Museum demonstrating light as narrative structure.

Across this continuum—from the analytical precision of the *Autochrome* to the emotional resonance of the *Scream Scaper*, the immersive spatiality of *VR CHROMA*, and the communicative synthesis of the *Gabinetto Segreto*—PERCEIVE progressively transformed color from an object of analysis into a medium of participation and care (Fig. 24).

4.2 A Tangible Demonstrator on Colour Change, Conservation and Care

Scream Time Machine is a demonstrator of the PERCEIVE project, designed within the Scenario 2 of the project (paintings and works on paper). It is conceived as an immersive experience of the iconic "The Scream" (1910?) painting owned by MUNCH museum in Oslo, in a multisensory approach. This experience was showcased and evaluated at the InArt24 pre-conference workshop (Colour and appearance for cultural heritage - new research approaches from the ApPEARS and PERCEIVE projects, 3rd June 2024) and at the Cultural History Museum (KHM) in Oslo on 8th June 2024.[1]

The proof of concept enhances the access and a sense of care for a fragile and precious artwork, deepens curiosity, and provides personal relevance and interpretation of the colored surface of the painting and its change over time, without requiring the download of an app. Both conservation and scientific perspectives around colour change (mainly fading) and interactive exploration with the surface and materiality of the object were considered in the design of the

[1] Scientific program - Museum of Cultural History, Oslo: https://www.khm.uio.no/english/research/inart2024/conference-programme/index.html#june-3.

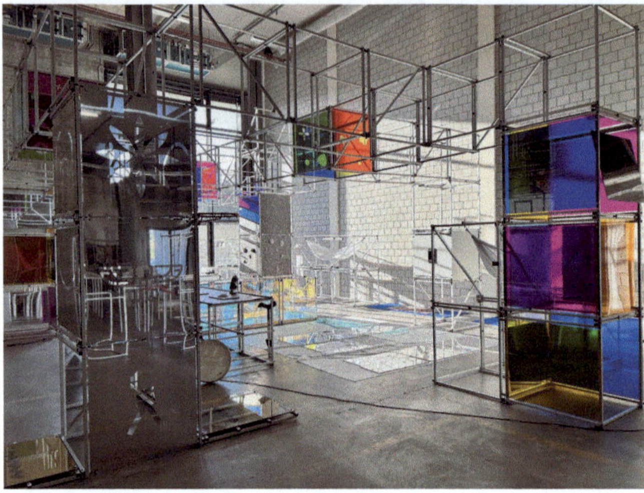

Fig. 24. The Open Space Museum—architectural synthesis of physical and digital color, light, and care.

different components of this experience. It is intended for use in both immersive and digital experiences or online contents.

The concept could be further developed into specific modules enhancing the different possible perceptions visitors could have: i.e. tactile (physical interaction with 3D prints and a 2D scan) or auditory and visual (through the translation of colours into sounds).

It consists of three components:

1. Video Tutorial (6.5 min, with subtitles in English)[2] that:
   - Provides context and background information about "The Scream" (1910?), its story and its scientific investigation spanning several years and different research projects (MOLAB [1,44,45], PERCEIVE [37,43] etc.).
   - Explains the various parts of the exploration including the production and background behind the acquisition of the 2D and 3D scans and the creation of the printed 3D replicas.
   - Introduces keywords and images used in the interactive Board Game.
2. Scream Reinvented Display:
   - Features a full 2D scan (30x) of "The Scream" (1910?) acquired in 2024 using high resolution Hirox digital microscopy technology6.
   - Includes several 3D prints (both monochrome and colored) created with various software and printing technologies, highlighting different areas of the painting affected by color change and its multilayered features [10,47].
3. Interactive Board Game:

[2] Full video tutorial, Fig. 25, can be found here: https://perceive-horizon.eu/index.php/results/perceive-tools/demonstrator-scream-time-machine/.

- Allows players to explore scientific concepts around color change (primarily fading) and conservation actions related to "The Scream" (1910?).
- Designed for a minimum of two players, where one can act as the Danger Master and the others as Time Travellers. Every time a Time traveller mistakes the date for the right sequence of the cards with an image of the Scream, the Danger master is entitled to stop the player and present a Danger card for which the appropriate Care card (action for mitigating the danger) has to be identified. The game can continue once the player chooses the right Care card for the Danger card. If the Danger Master plays all the Danger cards before the Time Travellers complete putting the cards in the correct order, they will lose the game. If they complete the timeline without making so many mistakes, they win.

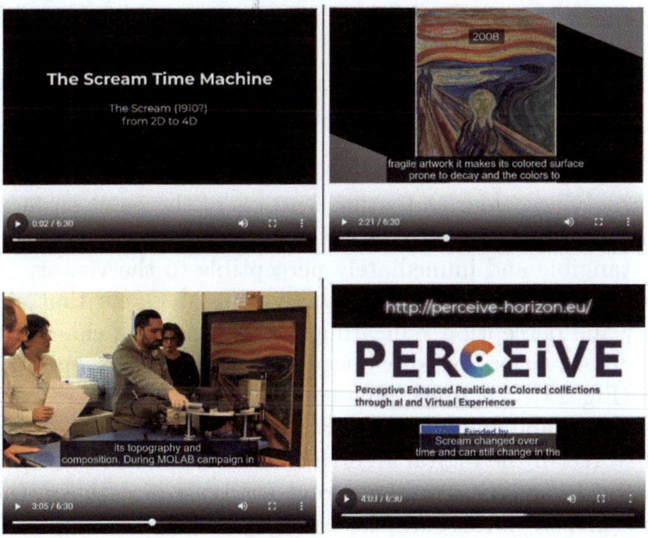

Fig. 25. Video frames from the tutorial explaining the story and research performed around the Scream

The Scream Time Machine experience begins with a video tutorial explaining the history around the photographic documentation and scientific investigation performed on the Scream since 1938 and to the present (Fig. 25).

Next, the public is introduced to the 2D scan (full size, at 30x magnification) and 3D prints made from Hirox scans of two specific areas of the painting (the screaming face and a sky detail), which are known to be affected by colour change, namely the fading of cadmium (Cd) yellow shades (Figs. 26 and 27).

These physical replicas consisting of 3D prints of details of the painting were created from high-resolution images captured with a Hirox digital microscope. These easy-to-understand physical replicas of parts of the painting can be

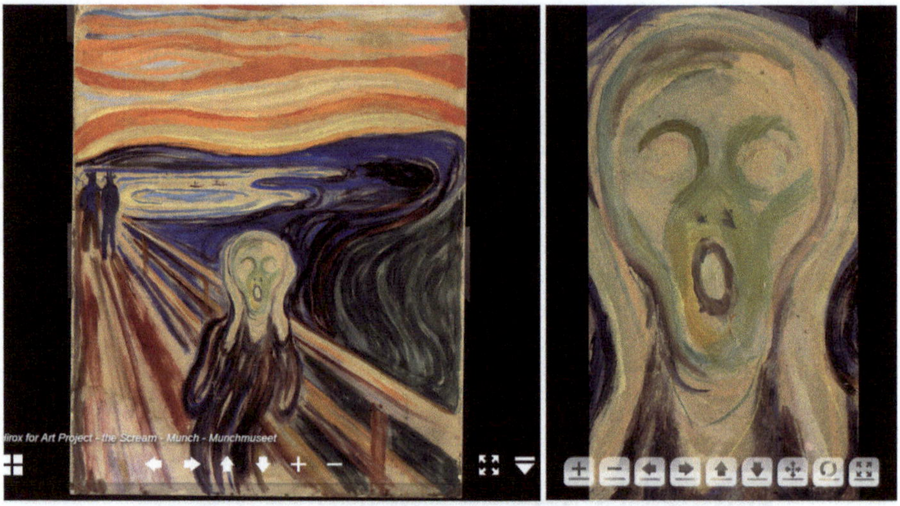

Fig. 26. Full size scan in 2D (30x) of the Scream and the face detail as a 3D scan (90x) (Color figure online)

touched. By resolving colour or texture features of the paint surface in different heights in the 3D prints, the colours of the painting as well as their change over time become tangible and immediately perceptible to the visitor.

The experience ends with the interactive Board Game that challenges the user's memory and perception, educating them about color change over time and the scientific conservation measures needed to prevent or stop the degradation phenomenon (Fig. 28). The dark blue cards of the Timeline represent 8 digital photographs, B&W and colored, of the Scream, taken in several years from 1938 to 2024, and stored in the digital archive of Munch museum, except the last one that is from the Hirox 2D acquisition and available directly on the website of Hirox Europe (https://hirox-europe.com/gigapixel/scream30x).

The game has been provided with a Manual and also an extended explanation of the Danger and Care cards terminology (Conservation vocabulary), in both English and Norwegian languages, as follows:

Dangers:

1. High Humidity: is the main cause of fading for Cadmium yellow pigment in the Scream palette and can also cause swelling and other types of damage to cardboard and other moisture-sensitive paint materials
2. Intense Light: radiation in the Visible and UV range can damage sensitive colors/pigments if too intense or exposed to it for prolonged periods (light-induced damage is cumulative)
3. Surface Damage: decay or degradation of the surface of the paint caused by mechanical agents or aggressive actions from the environment or during sudden events (robbery, vandalism, etc.)

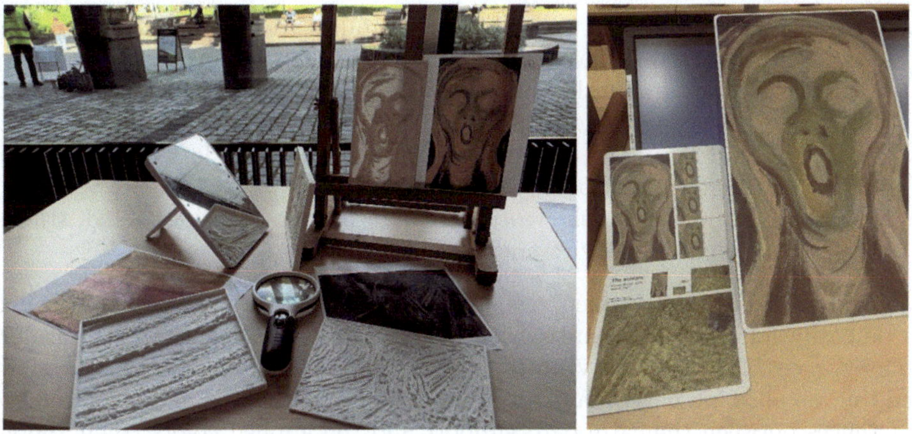

Fig. 27. Monochrome and colored 3D prints executed from the Hirox scan (Color figure online)

4. Paint Loss: loss and fading of a paint layers/materials due to different factors (aggression, abrasions, excessive exposure to low humidity and/or intense light that causes decohesion of paint and subsequent crumbling of the paint materials, etc.)
5. Robbery: a vandal action in which the art-object (painting) is stolen and removed from its place of display or storage

Care:

1. Stabilize: action meant to recover the stability of a paint/art-object if overexposed or underexposed to certain environmental parameters (light, relative humidity, pollution, etc.)
2. Monitor: action meant to measure and keep under stable levels the environmental parameters (light, relative humidity, pollution, etc.) that can potentially become damaging agents for a painting
3. Control exposure: action meant to control the levels of light exposure for exhibition/storage environments where fragile and sensitive art-objects are displayed
4. Map: action performed using a scientific equipment to identify and show the distribution of the materials present on a painted surface (this information can be further related to the technique of the artist and possible degradation or damage patterns on the surface)
5. Reconstruct: action aimed to build, re-establish, regenerate and/or simulate the appearance and colors of a painting surface if damaged or destroyed by external agents
6. Consolidate: action aiming to re-establish the cohesion between paint materials if there was a loss or degradation process affecting their integrity and stability

7. Rejuvenate: action meant to restore or revitalize the condition and characteristics of a painted surface (it can also mean reintegrating the paint colors if loss has occurred and also simulate the original appearance and colors if fading or other type of color change occurred over time);
8. Recover: action meant to recover a painting if stolen or lost due to several dangerous situations (robbery, vandalism, etc.)
9. Exhibit: action meant to showcase/display a painting in a protected, stable, and friendly accessible environment.

All three components (video, scans/prints, and the Board Game) are designed to be easily adapted to a multisensory immersive context, including considerations for visually impaired visitors. The experience engages two senses—sight and touch—allowing participants to explore surface patterns and understand the differences between the cardboard support and the applied colors (their topography and layered sequence), as well as between faded and non-faded areas of paint. Additionally, it explores its changes over time, some of which were due to unexpected events such as the 2004 robbery that damaged the left lower corner of the cardboard and also sensitive color shades on the bridge, as shown in the video. By integrating these elements, the Scream Time Machine offers a comprehensive, multisensory exploration of "The Scream (1910?)" and its conservation issues, enhancing the educational and experiential value for all visitors.

The visitor feedback of this experience during the InArt24 conference, was very positive. Tactile 3D-printed parts of "The Scream" painting were easy-to-understand and proved to be both engaging and fun. A 88.7% positive feedback was received about the 3D-printed tactile experience, the work was commented as being an innovative way to present scientific data. With the right guidance and information, the overall 3D prints experience can effectively engage and educate audiences.

The combination of high-resolution scans, 3D printed replicas, and interactive storytelling was evaluated positively for making complex scientific and conservation concepts accessible and engaging. Participants also appreciated the structured approach to explaining the lost colors, which provided a deeper appreciation of the artwork's original aesthetic and historical context. These insights highlight the importance of combining educational content with hands-on, multisensory experiences to foster a deeper and more memorable connection with cultural heritage [47].

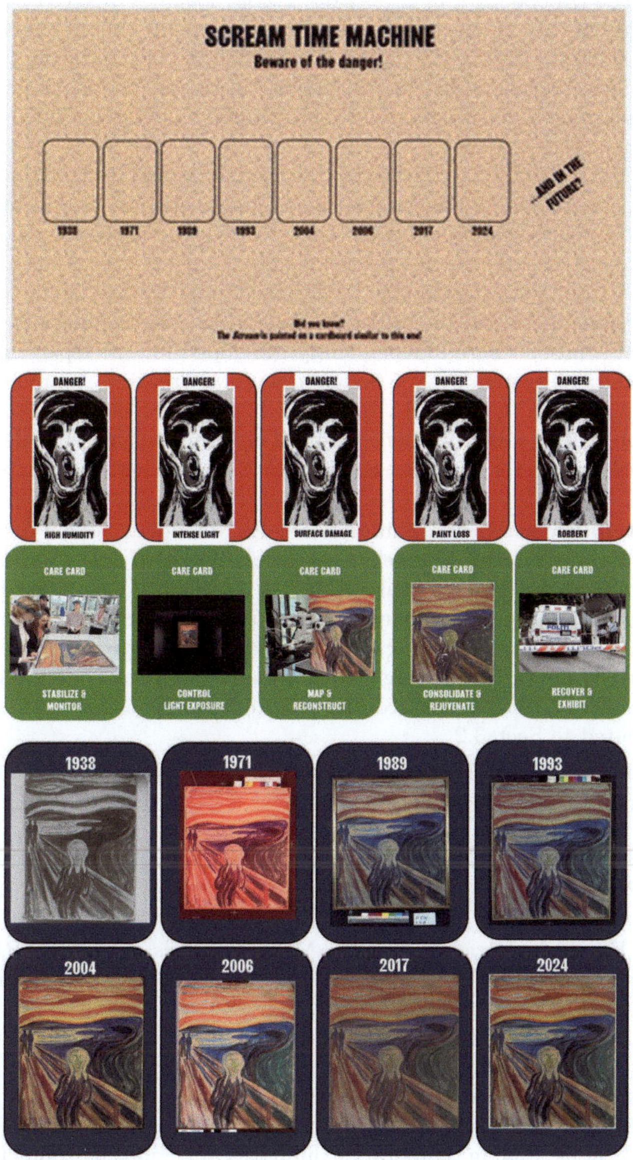

Fig. 28. The gameboard and the 3 sets of cards: 8 timeline cards (blue), 5 Danger cards (red) and 5 Care cards (green) (Color figure online)

Acknowledgments. This study was funded by the European Union (grant number 101061157, Project PERCEIVE). Research for the *CoDesign Toolbox* received additional funding from the projects H2IOSC (https://www.h2iosc.cnr.it/, funded by the European Union NextGenerationEU – National Recovery and Resilience Plan (NRRP) âĂŞ Project code IR0000029 âĂŞ CUP B63C22000730005), CHANGES (https://sites.google.com/uniroma1.it/changes, funded by the European Union NextGenerationEU - Project PE 0000020 CHANGES - CUP B53C22003780006) and from the Swiss Government Excellence Scholarship No. 2025.0089. Moreover, the CoDesign Tool builds upon the foundations of VisitorBox, developed under the Horizon 2020 project GIFT (Grant agreement ID: 727040).

The *Open Space Museum* project was conceptualized and designed by Arthur Clay, in collaboration with Dario Lanfranconi and Chiara Gemma Fedon at the Lucerne University of Applied Sciences and Arts (HSLU). Its structural logic draws inspiration from modular construction principles developed by Gerrit Rietveld [40] and Ken Isaacs [21], as well as from the open-source XYZ Spaceframe System created by N55 [31]. While these systems informed the project's underlying design philosophy, all structural and conceptual solutions were independently developed for the Open Space Museum. The XYZ Spaceframe system by N55 is shared under a Creative Commons Attribution-ShareAlike 3.0 License (CC BY-SA 3.0), which promotes democratic access, reuse, and adaptation in the spirit of open design. This ethos aligns with the Open Space Museum's mission to foster inclusive, participatory, and accessible cultural experiences.

Author contributions. Bruno Fanini, Marcello Massidda, Sofia Pescarin, Samuele Spotti, Laura Travaglini and Manuele Veggi contributed to Sect. 1; Cristiana Barandoni, Federica Bonifazi, Ivana Cerato, Enzo D'Annibale, Daniele Ferdani, Marialucia Giacco, Roberta Iannaccone, Donata Magrini, Caterina Serena Martucci, Samuele Spotti, Laura Travaglini, Sofia Pescarin and Manuele Veggi contributed to Sect. 2.1 and 2.2 (*Echoes of Loss*); Birgitte Aga and Nynne Ryhl Olsson contributed to Sect. 2.2 (*Tiny Conservators*); Arthur Clay contributed to Sect. 2.3 and 3.1; Irina Crina Anca Sandu contributed to Sect. 3.2

Disclosure of Interests. The authors have no competing interests to declare that are relevant to the content of this article.

References

1. Albertin, F., et al.: Iterative Fuzzy C-Means analysis of macro-XRF mapping of "The Scream" by Edvard Munch. In: MA-XRF scanning in Conservation, Art and Archaeology, MAXRF2022 (2022)
2. Almond, A.: Acqknowledge (Software), pp. 1–2. Wiley, Hoboken, November 2017
3. Aslan, S., et al.: Exploring kid space in the wild: a preliminary study of multimodal and immersive collaborative play-based learning experiences. Educ. Technol. Res. Dev. **70**(1), 205–230 (2022). https://doi.org/10.1007/s11423-021-10072-x
4. Balzani, R., et al.: Saving temporary exhibitions in virtual environments: the digital renaissance of Ulisse Aldrovandi – acquisition and digitisation of cultural heritage objects. Digi. Appl. Archaeol. Cult. Heritage **32**, e00309 (2024). https://doi.org/10.1016/j.daach.2023.e00309
5. Barandoni, C., Giacco, M., Martucc, C.S., Pescarin, S. (eds.): Perceive Isis' Colours. I Campi del Sapere – Valtrend Editore (2025)
6. Bebko, A.O., Troje, N.F.: bmlTUX: design and control of experiments in virtual reality and beyond. i-Perception **11**(4), 204166952093840 (2020). https://doi.org/10.1177/2041669520938400

7. Bonifazi, F., Veggi, M., Massidda, M., Ferdani, D., Pescarin, S.: Implementing curiosity hooks and caring practices in the reconstruction of lost polychromy: design prototypes for interactive experiences. In: Campana, S., et al. (eds.) Digital Heritage. The Eurographics Association (2025). https://doi.org/10.2312/dh.20253108
8. Brookes, J., Warburton, M., Alghadier, M., Mon-Williams, M., Mushtaq, F.: Studying human behavior with virtual reality: the Unity Experiment Framework. Behav. Res. Meth. **52**(2), 455–463 (2020). https://doi.org/10.3758/s13428-019-01242-0
9. Carroll, G.R.: Authenticity: Attribution, Value, and Meaning, pp. 1–13. Wiley, Hoboken, May 2015. https://doi.org/10.1002/9781118900772.etrds0020
10. Chan, E., Aambø, M., Godzimirska, M., Grimstad, I., Lojewski, T., Sandu, I.: Light-induced color changes on "the scream" versions in the Munch museum collection. MUNCH2022 Int. Conf. **13**(Special Issue 1), 1623–1634 (2022)
11. Clay, A.: Adaptive horizons: navigating museums in a post-pandemic world, November 2023. https://perceive-horizon.eu/index.php/2023/11/09/adaptive-horizons-navigating-museums-in-a-post-pandemic-world/
12. Clay, A.: A COVID-safe design approach for the Open Space Museum (OSM), February 2025. https://perceive-horizon.eu/index.php/2025/02/17/a-covid-safe-design-approach-for-the-open-space-museum-osm/
13. Commins, S., et al.: NavWell: a simplified virtual-reality platform for spatial navigation and memory experiments. Behav. Res. Meth. **52**(3), 1189–1207 (2020). https://doi.org/10.3758/s13428-019-01310-5
14. De Caro, S.: Il santuario di Iside a Pompei e nel Museo archeologico nazionale. Electa Napoli/Soprintendenza per i beni archeologici di Napoli e Caserta, Napoli (2006)
15. Falvo, P.G.: In the mirror of the wonder of Luca Giordano: multisensory experience itinerary with impact assessment at Palazzo Medici Riccardi (Florence, Italy). In: 2012 18th International Conference on Virtual Systems and Multimedia, pp. 11–18 (2012). https://doi.org/10.1109/VSMM.2012.6365901
16. Ferdani, D., Barandoni, C., Bonifazi, F., Iannaccone, R., Magrini, D.: Multiband photogrammetry for multispectral 3D reconstruction of statuary. J. Cult. Herit. **76**, 74–85 (2025)
17. Floridi, L.: The Fourth Revolution: How the Infosphere is Reshaping Human Reality. Oxford University Press, UK (2016)
18. Fogg, B.J.: Persuasive technology. Ubiquity **2002**(December), 2 (2002)
19. GIFT: Meaningful Personalization of Hybrid Virtual Museum Experiences Through Gifting and Appropriation (2023). European Commission, CORDIS. https://doi.org/10.3030/727040
20. International Council of Museums (ICOM): ICOM voices: sustainability management in museums, May 2023. https://icom.museum/en/news/icom-voices-sustainability-management-museums/
21. Isaacs, K.: How to Build Your Own Living Structures. Harmony Books (1974)
22. Khaled, R., Vasalou, A.: Bridging serious games and participatory design. Int. J. Child Comput. Interact. **2**(2), 93–100 (2014). https://doi.org/10.1016/j.ijcci.2014.03.001
23. Kidd, J.: Museums in the New Mediascape. Digital Research in the Arts and Humanities. Ashgate Publishing, London, England, September 2014
24. Li, H., Zhang, M.: Museum game-based learning: innovative approaches from a constructivist perspective. Front. Educ. **10**, 1576207 (2025). https://doi.org/10.3389/feduc.2025.1576207

25. Li, Y., Said, T.S.B.: Children's learning experiences in immersive museum art education: an observational case study. Int. J. Educ. Humanit. **19**(3), 165–169 (2025). https://doi.org/10.54097/pdn2r442
26. Loudoun, F.M., Boyle, B., Larsson-Lund, M.: Play value of digital play spaces: children's voices. Int. J. Child Comput. Interact. **40**, 100649 (2024). https://doi.org/10.1016/j.ijcci.2024.100649
27. Mairesse, F.: Dictionary of Museology, 1 edn. Routledge, London, April 2023. https://doi.org/10.4324/9781003206040
28. Massidda, M., et al.: Towards an experiment planner for cognitive studies in virtual heritage environments. A pilot study. In: Corsini, M., Ferdani, D., Kuijper, A., Kutlu, H. (eds.) Eurographics Workshop on Graphics and Cultural Heritage. The Eurographics Association (2024). https://doi.org/10.2312/gch.20241261
29. Massidda, M., Fanini, B., Pescarin, S.: Prototyper: a web3D platform for collaborative design and simulation of hybrid museum exhibitions. In: Campana, S., et al. (eds.) Digital Heritage. The Eurographics Association (2025). https://doi.org/10.2312/dh.20253009
30. Muravevskaia, E., Gardner-McCune, C.: Designing a virtual reality empathy game framework to create empathic experiences for children. Int. J. Child Comput. Interact. **35**, 100561 (2023). https://doi.org/10.1016/j.ijcci.2022.100561
31. N55: XYZ Spaceframe Vehicles. N55 Publications, Copenhagen, Denmark (2003)
32. open.museum: Perceive demonstrators: autochrome, scream scaper, VR chroma, and Gabinetto Segreto (2025). https://openspacemuseum.blogspot.com/p/demonstrators.html
33. Papoutsi, C., Drigas, A., Skianis, C., Skordoulis, C., Pappas, M.: "The park of emotions": a serious game for the development of emotional intelligence in children. Appl. Sci. (Basel) **14**(14), 6067 (2024). https://doi.org/10.3390/app14146067
34. PERCEIVE Project: Arthur clay's interview for perceive parcour exhibition (2024). https://perceive-horizon.eu/index.php/2024/11/29/arthur-clays-interview-for-perceive-parcour-exhibition
35. PERCEIVE Project: New paths for ideation: the open space museum sketch book (2024). https://perceive-horizon.eu/index.php/2024/09/19/new-paths-for-ideation-the-open-space-museum-sketch-book/
36. PERCEIVE Project: Toward creating an autochrome demonstrator for museums (2024). https://perceive-horizon.eu/index.php/2024/05/30/toward-creating-an-autochrome-demonstrator-for-museums/
37. Pescarin, S., et al.: Perceptive enhanced realities of coloured collections through AI and virtual experiences. In: Campana, S., et al. (eds.) DIGITAL HERITAGE 2025, Siena, Italy, September 2025, pp. 350. Eurographics - The European Association for Computer Graphics (2025)
38. Pescarin, S., et al.: Authenticity in the design of digital heritage applications. In: Chromatic Visions. Exploring Colour in Art, Archaeology and Digital Realities, Part I. LNCS, pp. 81–127. Springer, Cham (2026). https://doi.org/10.1007/978-3-032-07792-9_3
39. Pescarin, S., Città, G., Spotti, S.: Authenticity in interactive experiences. Heritage **7**(11), 6213–6242 (2024). https://doi.org/10.3390/heritage7110292
40. Rietveld, G., Vöge, P., Overy, P.: The Complete Rietveld Furniture. Monografieën van Nederlandse architecten. 010 Publishers (1993)
41. Rubin, R.B., Rubin, A.M., Graham, E., Perse, E.M., Seibold, D.: Self-assessment manikin. In: Communication Research Measures II, pp. 354–359. Routledge (2010)

42. Saavedra, A., Moreno-Lobato, A., Costa-Feito, A.: Application and development of emotional and attentional tools: a face reading, eye tracking, and electrodermal activity approach. In: Liebana-Cabanillas, F., Muñoz-Leiva, F., Prados-Castillo, J., Higueras-Castillo, E. (eds.) Biometrics. CRC Press/Taylor & Francis (2025). https://doi.org/10.1201/9781003509554-4
43. Sandu, I., Ciortan, I.: Perceive: a new approach to the simulation of color change in paintings and works on paper. In: Online webinar organized by ICOM-CC (Paintings, Paper and Scientific Research groups, held on 10th and 21st March 2025 (2025)
44. Sandu, I., et al.: The "scream" by Edvard Munch - same motif, different colours, different techniques and approaches – a novel non-invasive comparative study of two of its versions. In: CHEM CH Conference, oral presentation, July 2018, Bucharest (2018)
45. Sandu, I., et al.: TECH_SCREAM_TECH – TECHnical Study of CoRE Art Materials & TECHniques of SCREAM versions. Poster presented at IPERION CH 3rd Transnational Access Users meeting: c2RMF. Palais du Louvre, Paris, France (2019)
46. Simon, N.: The Participatory Museum. Museum 2.0 (2010)
47. Stavroulakis, P., et al.: Tactile 3D-printed media interaction with the color surface features of paintings. the case of the Scream (1910?) painting. Front. Comput. Sci. **7**, 1597880 (2025). https://doi.org/10.3389/fcomp.2025.1597880
48. Stein, J., Adams, M., Luke, J.: Thinking evaluatively: a practical guide to integrating the visitor voice. Hist. News **62**(2), 1–8 (2007)
49. Stolterman, E., Wiberg, M.: Concept-driven interaction design research. Hum. Comput. Interact. **25**(2), 95–118 (2010)
50. Travaglini, L., Massidda, M., Zapf, A., Spotti, S., Veggi, M., Pescarin, S.: The codesign tool. A card-based framework for interactive and collaborative design in cultural heritage. In: Schrepp, M., Rauterberg, M. (eds.) HCI International 2025 - Late Breaking Papers: Part XII. Lecture Notes in Computer Science, vol. 16342. Springer, Cham (2026). https://doi.org/10.1007/978-3-032-13164-5_18
51. Veggi, M., et al.: Sense of care in the design of digital heritage applications. In: Chromatic Visions. Exploring Colour in Art, Archaeology and Digital Realities, Part I. LNCS, pp. 128–161. Springer, Cham (2026). https://doi.org/10.1007/978-3-032-07792-9_4
52. Veggi, M., von Gal, A., Piccardi, L., Pescarin, S., Nori, R.: How much do we care about cultural heritage? A Rasch scale validation study among young adults. Heritage **8**(8), 317 (2025)
53. Xavier, A., et al.: Children's perspectives on using serious games as a complement to promoting their social-emotional skills. Int. J. Environ. Res. Pub. Health **19**(15), 9613 (2022). https://doi.org/10.3390/ijerph19159613
54. van Zomeren, M., Saguy, T., Schellhaas, F.M.H.: Believing in "making a difference" to collective efforts: participative efficacy beliefs as a unique predictor of collective action. Group Process. Intergroup Relat. **16**(5), 618–634 (2013)

Author Index

I. M. Ciortan et al. (eds.), Chromatic Visions, LNCS 16450, p. 147, 2026.
https://doi.org/10.1007/978-3-032-17700-1

MIX
Papier aus verantwortungsvollen Quellen
Paper from responsible sources
FSC® C105338

FSC
www.fsc.org

If you have any concerns about our products,
you can contact us on
ProductSafety@springernature.com

In case Publisher is established outside the EU,
the EU authorized representative is:
Springer Nature Customer Service Center GmbH
Europaplatz 3, 69115 Heidelberg, Germany

Printed by Libri Plureos GmbH
in Hamburg, Germany